©2001 Algrove Publishing Limited
ALL RIGHTS RESERVED.
No part of this book may be reproduced in any form, including photocopying, without permission in writing from the publishers, except by a reviewer who may quote brief passages in a magazine or newspaper or on radio or television.

Algrove Publishing Limited
36 Mill Street, P.O. Box 1238
Almonte, Ontario, Canada K0A 1A0

Telephone: (613) 256-0350
Fax: (613) 256-0360
Email: sales@algrove.com
www.algrove.com

Library and Archives Canada Cataloguing in Publication

Graham, Frank D. (Frank Duncan), b. 1875
   How to use the steel square

(Classic reprint series)
Originally published 1923 as chapter 23 of Audel's carpenters' and builders' guide #1.
Includes index.
ISBN-13 978-1-894572-34-7

1. Carpenters' squares. I. Emery, Thomas J. II. Graham, Frank D. (Frank Duncan), b. 1875. Audel's carpenters' and builders' guide #1. III. Title. IV. Series: Classic reprint series (Almonte, Ont.)

TJ5619.G73 2001   694'.2'028   C2001-900367-6

Printed in Canada
#8-08-10

Publisher's Note

There have been many texts written on the use of the steel square but most take the subject far beyond the interest of the regular user. The Audel's Carpenters Guides were extensive yet eminently practical. This section from the 1923 edition of *Audel's Carpenters And Builders Guide #1* is both practical and well illustrated, making it a valuable reference for anyone involved in frame construction.

> Leonard G. Lee, Publisher
> Ottawa
> March, 2001

How to Use
The Steel Square

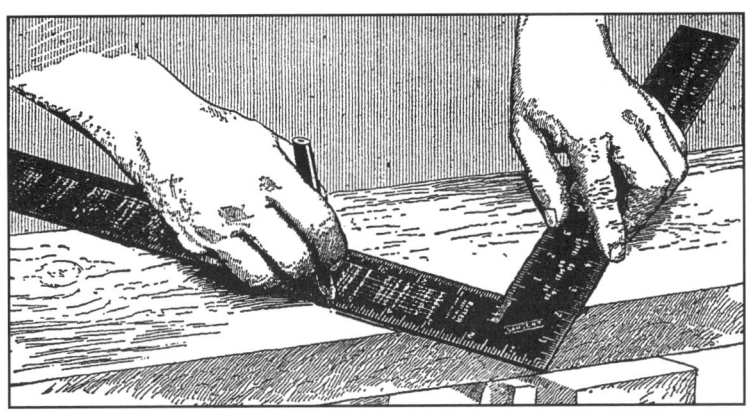

Originally published in
Audel's Carpenters and
Builders Guide #1 (1923)

Algrove Publishing
Classic Reprint Series

How to Use the Steel Square
Table of Contents

Parts of Framing Square 1
Markings of Square 3
Application of the Square 3
Scale Problems 3
Angle Table for Square 9
Square and Bevel Problems 10
    Main or Common Rafters 17
    Hip (and Valley) Rafters 22
    Valley Rafters 23
    Jack Rafters 24
    Cripple Rafters 24
    Finding Rafter Lengths Without
    Aid of Tables 24
Rafter Tables 29
    Class I 30
        Common Rafter Table 30
        Hip Rafter Table 32
    Class II - Reading length of rafter
    foot per run 38
    Table of Octagon Rafters 42
    Table of Brace Measure 45
    Octagon Table or Eight Square Scale 46
    Essex Board Measure Table 48
Index 49

How to Use The Steel Square

On most construction work, especially in house framing the so-called "steel square" is invaluable for accurate measuring and for determining angles. The author seriously objects to the term *"steel* square." The proper name is *framing square*, because the square with its markings was designed especially for marking timber in framing. However, the wrong name has become so firmly rooted that it will have to be put up with.

The square as a tool with its various scales and tables has been explained in Chapter 7. The present treatment is to explain more in detail these markings and their application by examples showing actual use of the square. The following names used to identify the different portions of the square should be noted and remembered:

Parts of Framing Square

Body.—The longer and wider member.

Face.—The sides visible (both body and tongue) when the square is held by the tongue in the right hand, the body pointing to the left.

Tongue.—The shorter and narrower member.

Back.—The sides visible (both body and tongue) when the square is held by the tongue with the left hand, the body pointing to the right.

The size square most generally employed is that with a 12

in. tongue and an 18 in. body, but there are many purposes which require a 16 to 18 in. tongue, and a 24 in. body.

The body of the larger is 2 in. wide and the tongue $1^3/_4$ ins. wide, $^3/_{16}$ in. thick at the heel or corner for strength, diminishing, for lightness to the two extremities to about $^3/_{32}$ in.

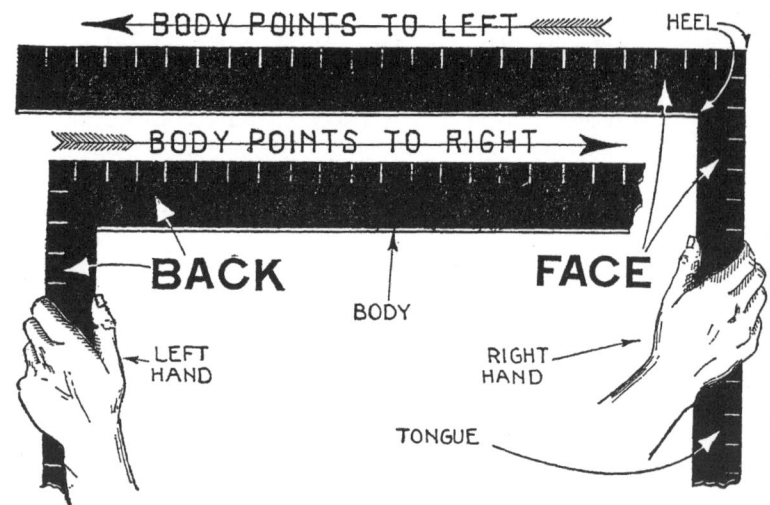

FIGS. 935 and 936.—Face and back sides of square with names used to identify its different portions. These are defined in the text and should be firmly fixed in mind by aid of the illustrations. The body of the square is sometimes called the *blade*.

The various markings on squares are of two kinds:

1. Scales or graduations.
2. Tables.

In buying a square it is advisable to get one with all the markings rather than a cheap square with some of the scales and tables omitted. Thus

Markings of Cheap Square

Tables   **Graduations**

Rafter—Essex—brace $1/16$, $1/12$, $1/8$, $1/4$

Complete Markings

Rafter, Essex, brace, octagon, $1/100$, $1/32$, $1/16$, $1/12$, $1/10$, $1/8$, $1/4$
polygon cuts

The square with the complete markings will cost more, but in the purchase of tools make it a rule *to purchase only the finest made*. The general arrangement of the markings on squares differ somewhat with different makes, hence, it is advisable to examine the different makes before purchasing to select the one best suited to individual requirements.

Application of the Square.—As before stated the markings on squares of different makes sometimes vary both in their position on the square and the mode of application, but a thorough understanding of the application of the markings on any first class square will enable the student to easily acquire proficiency with any other square.

The various markings may be divided into two groups:

1. Scales.
2. Tables.

The application of the scales will be first considered, as before explaining the use of the tables, the student should understand the general arrangement of roof frame work, names of the different kinds of rafters, other parts, etc.

Scale Problems.—The term *scales* is used to denote the inch divisions of the tongue and body length found on the outer and inner edges, and the inch graduations into $1/4$, $1/8$, $1/10$, $1/12$, $1/16$, $1/32$, and $1/100$. As before stated all these graduations should be on a first class square (hence, look for them in purchasing a

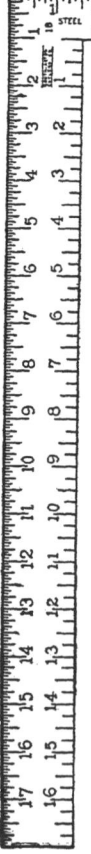

square)—but, on cheap squares will be found only a few of these graduations—as 1/16, 1/8, 1/4.

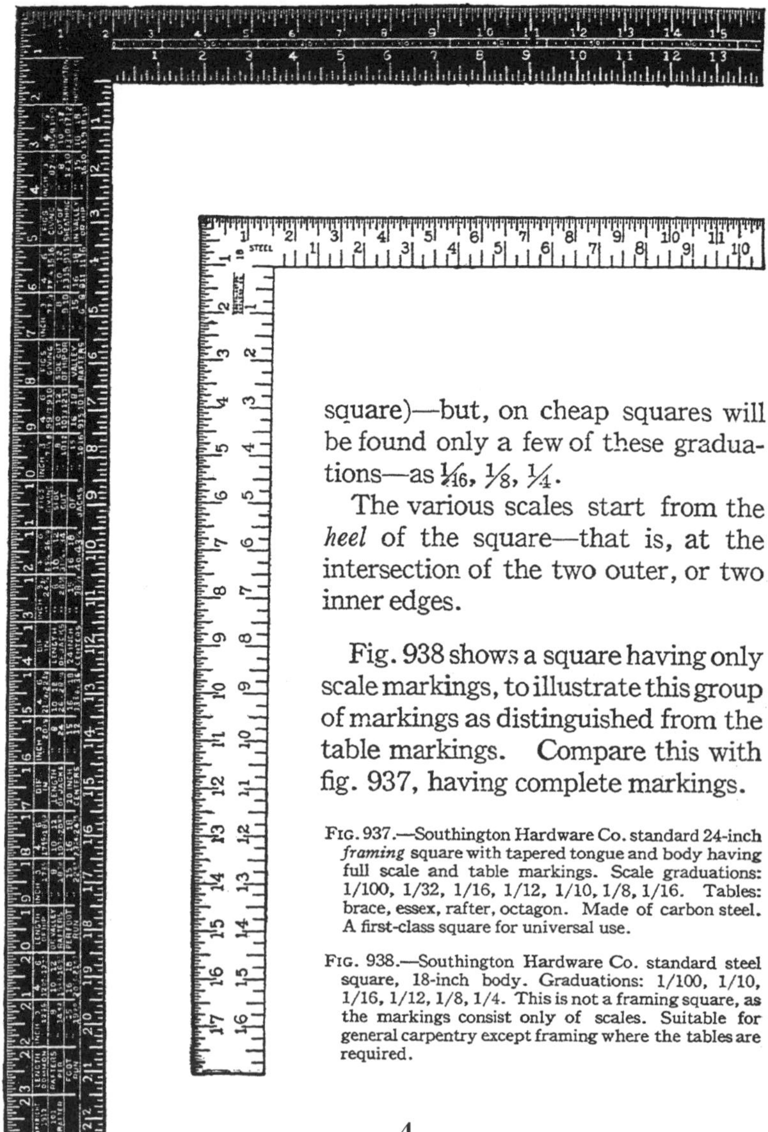

The various scales start from the *heel* of the square—that is, at the intersection of the two outer, or two inner edges.

Fig. 938 shows a square having only scale markings, to illustrate this group of markings as distinguished from the table markings. Compare this with fig. 937, having complete markings.

Fig. 937.—Southington Hardware Co. standard 24-inch *framing* square with tapered tongue and body having full scale and table markings. Scale graduations: 1/100, 1/32, 1/16, 1/12, 1/10, 1/8, 1/16. Tables: brace, essex, rafter, octagon. Made of carbon steel. A first-class square for universal use.

Fig. 938.—Southington Hardware Co. standard steel square, 18-inch body. Graduations: 1/100, 1/10, 1/16, 1/12, 1/8, 1/4. This is not a framing square, as the markings consist only of scales. Suitable for general carpentry except framing where the tables are required.

How to Use the Steel Square

A square having only the scale markings as shown in fig. 938, is adequate to solve many problems in laying out carpentry work. An idea of its range of usefulness is shown in the following problems.

Problem 1.—To describe a semi-circle with given diameter.

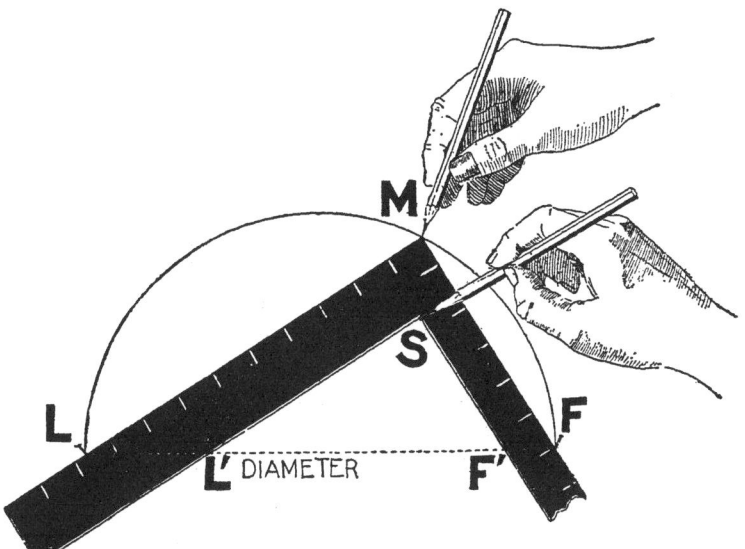

FIG. 939.—*Problem 1: To describe a semi-circle with given diameter.* **Outer heel method:** Drive brads at points L, F, extremities of the given diameter. With pencil held at the outer heel M, slide square around with its sides in contact with L, and F, then with the pencil held at M, describe a semi-circle. *Inner heel method:* Obviously if the pencil be held at S, it will be better guided, than at M. In this method, the distance L'F', should be taken to equal diameter, the inner edges of the square sliding on the tacks—the same edges (in either case) that guide the pencil.

At the ends of the diameter LF (fig. 939) drive brads. Place the outer edges of the square against the nails and hold a lead pencil at the outer heel M, any semi-circle can be described as indicated.

This is the *outer heel* method, but a better guide for the pencil is obtained by the *inner heel* method also shown in the figure.

How to Use the Steel Square

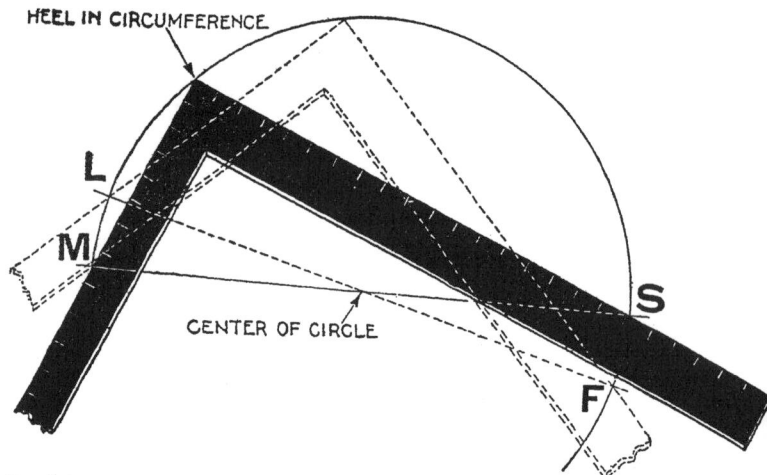

FIG. 940.—*Problem 2: To find the center of a circle.* At the points MS, and LF, where the sides of the square cut the circle when placed in any position with heel in circumference, draw diameter and then intersection will be the center of the circle. *Why?*

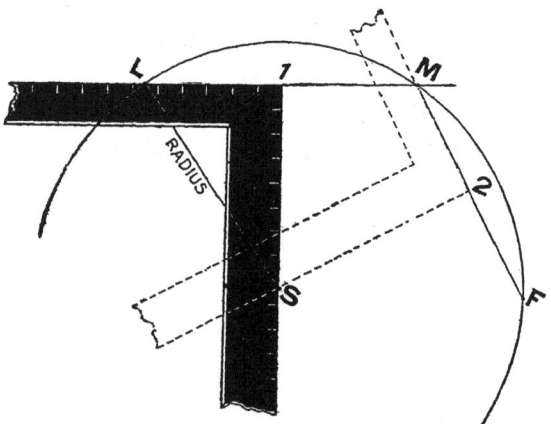

FIG. 941.—*Problem 3: To describe a circle through three points not in a straight line.* Let L, M, and F, be the given points. Join these points with lines LM, and MF, bisecting them at 1 and 2. Apply square with heel at 1 and, 2 as shown and the intersection of perpendiculars thus obtained at S, will be the center of circle which, with radius LS, may be described through LM and F.

How to Use the Steel Square

Problem 2.—*To find the center of a circle.*

Lay the square on the circle so that its outer heel lies in the circumference.

Mark the intersections of the body and tongue with the circumference. A line connecting these two points is a diameter and by drawing another diameter (obtained in the same way) the intersection of the two diameters is the center of the circle as shown in fig. 940.

Problem 3.—*To describe a circle through three points not in a straight line.*

Joint points with straight lines; bisect these lines and at the points of bisection erect perpendiculars with the square. The intersection of these

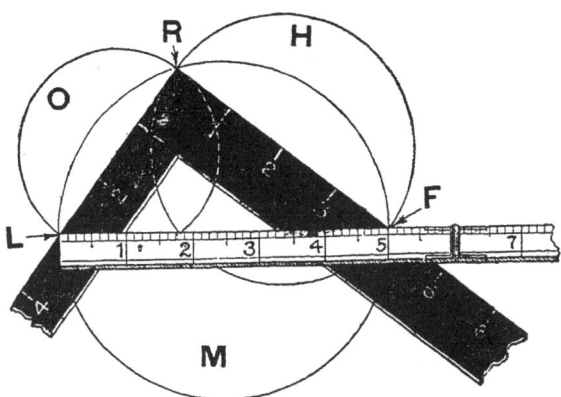

FIG. 942.—*Problem 4: To find the diameter of a circle whose area is equal to the sum of the areas of two given circles.* Let O, and H, be the given circles (drawn with diameters LR, and RF, at right angles). Suppose diameter of O, be 3 inches, and diameter of H, 4 inches. Then points L, F, at these distances from the heel of the square will be 5 inches apart as conveniently measured with a two-foot rule as shown. This distance LF, or 5 inches, is diameter of the required circle. *Proof:* $LF^2 = LR^2 + RF^2$, that is $5^2 = 3^2 + 4^2$ or $25 = 9 + 16$.

perpendiculars is the center from which a circle may be described through the three points as in fig. 941.

Problem 4.—*To find the diameter of a circle whose area is equal to the sum of the areas of two given circles.*

Lay off on tongue of square diameter of one of the given circles, and on body diameter of the other. The distance between these points (measure across with a two foot rule) will be diameter of the required circle as in fig. 942.

Problem 5.—*To lay off angles of* 30° *and* 60°.

Mark off 15 ins. on a straight line and lay the square so that the body

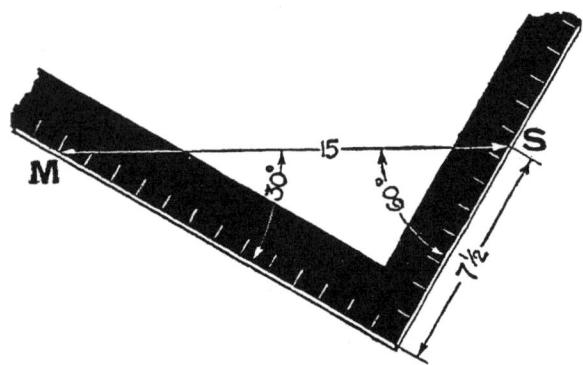

FIG. 943.—***Problem 5:*** *To lay off angles of* 30° *and* 60°. Draw line MS, 15 inches long. Place square so that S, touches tongue 7½ inches from hip, and M, touches body. The triangle thus formed will have an angle of 30° at M, and 60 at S.

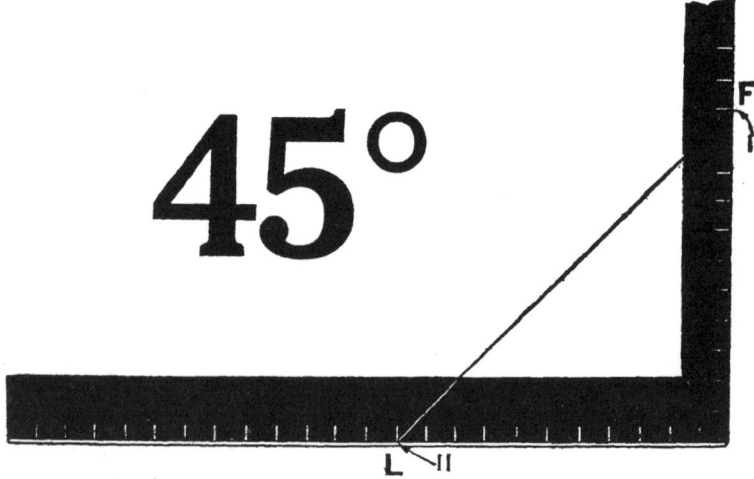

FIG. 944.—***Problem 6:*** To lay off an angle of 45°. Take equal measurements L, F, on body and tongue of the square then with ∠L = ∠F = 45°.

How to Use the Steel Square

touches one mark and 7½ ins. on the tongue is against the other mark as in fig. 943. The tongue will then form an angle of 60° with the line, and the body, 30°.

Problem 6.—*To lay off an angle of 45°.*

The diagonal line connecting equal measurements on either arm of the square forms angles of 45° with the blade and tongue as in fig. 944.

Problem 7.—*To lay off any angle.*

The accompanying table gives values for measurements on tongue and body of the square such that by joining the points corresponding to the measurements any angle may be laid out from 1 to 45° as explained in fig. 945.

Angle Table for Square

Angle	Tongue	Body	Angle	Tongue	Body	Angle	Tongue	Body
1	.35	20.	16	5.51	19.23	31	10.28	17.14
2	.7	19.99	17	5.85	19.13	32	10.6	16.96
3	1.05	19.97	18	6.58	19.02	33	10.89	16.77
4	1.4	19.95	19	6.51	18.91	34	11.18	16.58
5	1.74	19.92	20	6.84	18.79	35	11.47	16.38
6	2.09	19.89	21	7.17	18.67	36	11.76	16.18
7	2.44	19.85	22	7.49	18.54	37	12.04	15.98
8	2.78	19.81	23	7.8	18.4	38	12.31	15.76
9	3.13	19.75	24	8.13	18.27	39	12.59	15.54
10	3.47	19.7	25	8.45	18.13	40	12.87	15.32
11	3.82	19.63	26	8.77	17.98	41	13.12	15.09
12	4.16	19.56	27	9.08	17.82	42	13.38	14.89
13	4.5	19.49	28	9.39	17.66	43	13.64	14.63
14	4.84	19.41	29	9.7	17.49	44	13.89	14.39
15	5.18	19.32	30	10.	17.32	45	14.14	14.14

Problem 8.—*To find the octagon of any size square timber.*

Place the body of a 24 in. square diagonally across the timber so that both extremities (ends) of the 24″ body touch opposite edges. Make a mark at 7 ins. and 17 ins. as in fig. 946. Repeat the process at the other end and draw lines through the pairs of marks, these lines showing the portion of material necessary to come off the corners.

How to Use the Steele Square

Square and Bevel Problems.—By the application of a large bevel to the framing square, it becomes a calculating machine, and by its means arithmetical processes are greatly simplified. This bevel is preferably made of steel blades, procurable from a tool maker; the following points being observed in its construction:

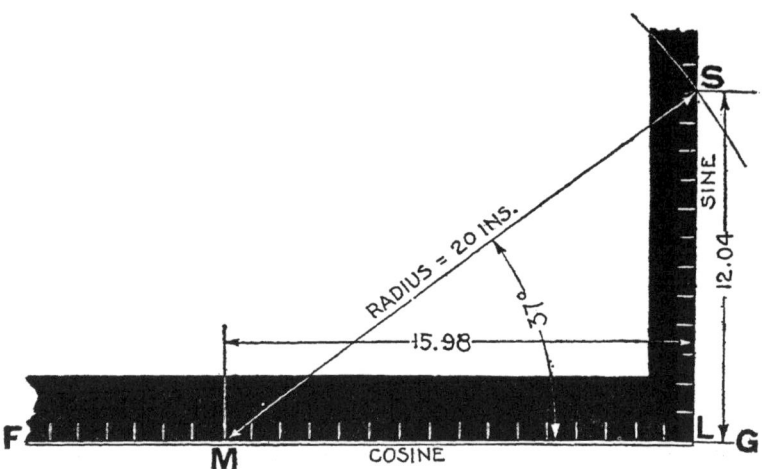

FIG. 945.—*Problem 7: To lay out any angle.* Let 37° be the required angle. Place body of square on the line FG, and from the table lay off on tongue LS = 12.04 inches, and LM, on body = 15.98 inches. Draw MS, then angle LMS = 37°. By measurement MS, will be found to be equal to 20 inches for any angle, because the values given in the table for LS, and MS, are *natural sines* and *natural cosines* multiplied by 20, hence MS = 1 ×20.

The edges of each blade must be true; the blade *e* in fig 948 must lie under the square so as not to hide the graduations; the

NOTE.—The side of an inscribed octagon can be obtained from the side of a given square, by multiplying the side of the square by five and dividing the product by twelve. The quotient will be the side of the octagon.

NOTE.—The side of a *hexagon* is equal to the radius of the circumscribing circle. If the side of a desired hexagon be given, arcs should be struck from each extremity of it at a radius equal to its length. The point where these arcs intersect is the center of the circumscribing circle, and having described it, it is sufficient to prick off chords on its circumference, equal to the given side, to complete the hexagon.

How to Use the Steel Square

two blades must be fastened together by a thumb screw to lock them; the blade *l* should have a hole near each end and one in the middle so that blade *e* may be shifted as required, with a

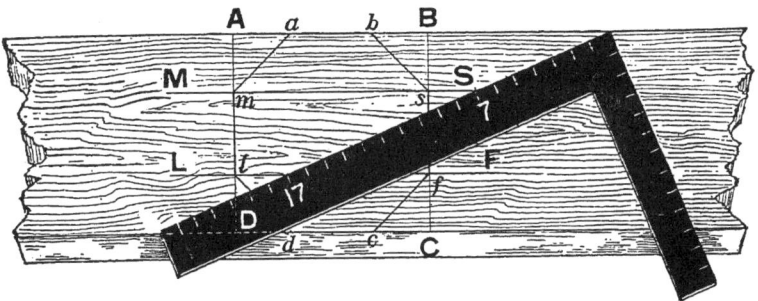

FIG. 946.—*Problem 8: To find the octagon of any size timber.* First lay out a square ABCD. Placing the body of a 24-inch square as shown parallel lines MS and LF, are drawn through points 7 and 17 as shown. These intercept sides *ml*, and *sf*, of the octagon. To lay off side *sb*, place square so that tongue touches *s*, and body touches *l*, with heel touching line AB. The remaining sides are obtained in a similar manner.

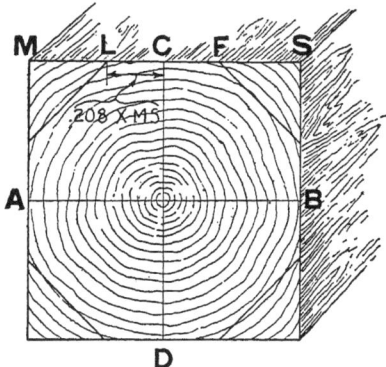

FIG. 947.—*Problem 8: To find the octagon of any size timber* (second method). Let AB, and CD, be center lines and MS, one side of the square timber. *Rule: Multiply length of side by .208 and product is half side of octagon.* Thus lay off CF and CL, each = .208 ×MS, then LF, is side of octagon. Set dividers to distance CL and lay off other sides from centers A,B,D, and complete polygon.

large notch as shown, near each hole in order to observe the position of blade *e*.

Problem 9.—*To find the diagonal of a square.*

Set the blade *e* to 10⅜ on the tongue and 15 on the body. Assume an 8 in. square. Slide the bevel sidewise along the tongue until the blade *e*, is against 8, when the other edge will touch 11⁵⁄₁₆ on the body which is the required diagonal.

Problem 10.—*To find the circumference of a circle from its diameter.*

Set the bevel blade to 7 on the tongue of the square and to 22 on the body. The reading on the body will be the circumference corresponding to the

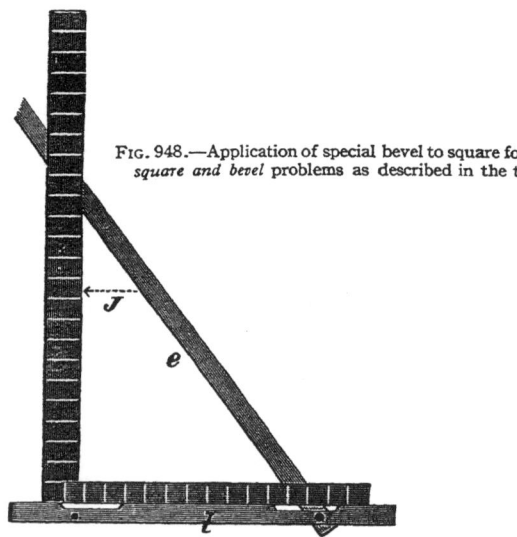

Fig. 948.—Application of special bevel to square for solving *square and bevel* problems as described in the text.

diameter to which *e*, is set upon the tongue. To reverse the process, use the same bevel, and read the required diameter from the tongue, the circumference being set on the body

Problem 11.—*Given the diameter of a circle, to find the side of a square of equal area.*

Set the bevel blade to 10⅝ on the tongue and 12 on the body, then the diameter of the circle, on the body, will give the side of the equal square upon

How to Use the Steel Square

the tongue. If the circumference be given instead of the diameter, set the bevel to 5½ on the tongue and 19½ on the body, finding the side of the square on the tongue as before.

Problem 12.—*Given the side of the square, to find the diameter of a circle of equal area.*

This, together with the preceding problem, is very useful in making calculations for spouts and pipes. Using the same bevel as in Problem 11, the blade *e*, is set to the given side upon the tongue of the square, the required diameter being read off the body.

Problem 13.—*Given the diameter of the pitch circle of a gear wheel, and the number of teeth; to find the pitch.*

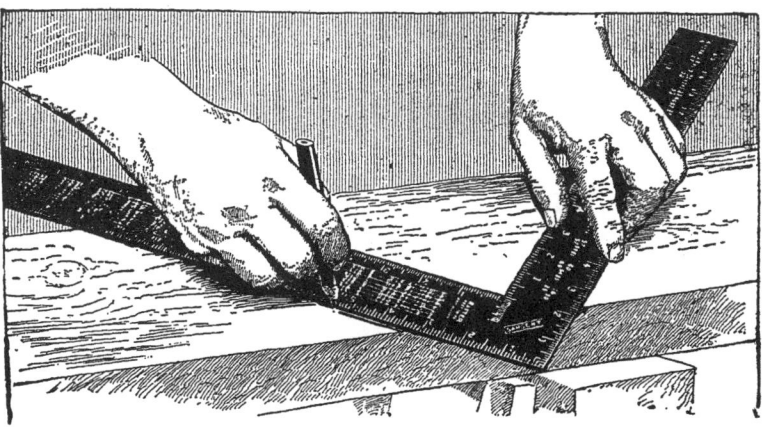

FIG. 949.—Framing square in use on board.

Take the number of teeth or a proportional part upon the body of the square, and the diameter or a similar proportional part upon the tongue, and set the bevel blade to those marks. Slide the bevel along to 3.14 on the body, and the number given on the tongue, multiplied by the proportional divisor will be the required pitch.

Problem 14.—*Given the pitch of teeth and diameter of pitch circle in a gear wheel, to find the number of teeth.*

Set the bevel blade to the pitch on the tongue, and 3.14 on the body of the square. Move the bevel along until it marks the diameter upon the tongue when the number of teeth can be read from the blade. If the

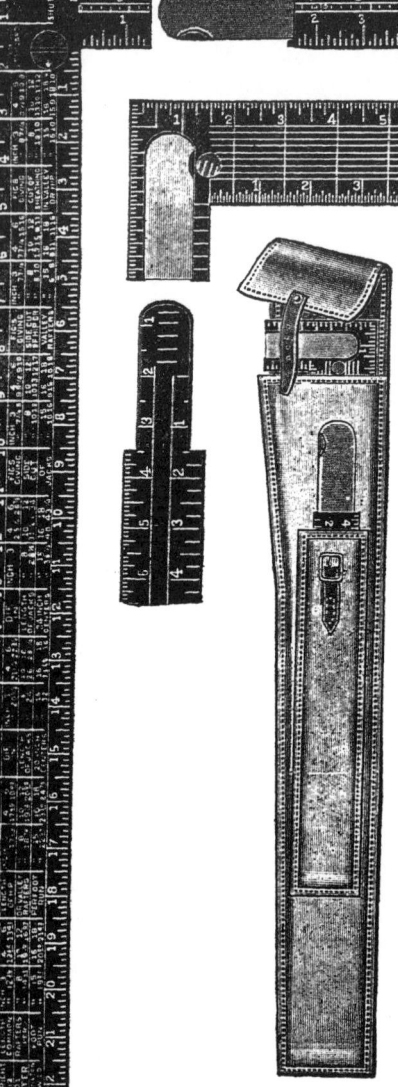

FIGS. 950 to 954.—Southington standard take-down square. Fig. 950, face of body; fig. 951, face of tongue; fig. 952, back of body; fig. 953, back of tongue; fig. 954, square packed in scabbard. The square has a one-piece solid heel. The tongue fits easily and locks with an anchored cam. The cam lock may be turned by a screw driver or coin. The long bearing joint gives maximum strength and insures the truth of the square.

How to Use the Steel Square

diameter be too large for the tongue, divide it into proportional parts, also the pitch, multiplying the number found by the same figure.

Problem 15.—*The side of a polygon being given to find the radius of the circumscribing circle.*

Set bevel to the pairs of numbers in the table below taking one-eighth or one-tenth of an inch as a unit. The bevel, when locked, is slid along to the given length of side, and the required length of radius is read upon the other leg of the square.

TABLE FOR INSCRIBED POLYGONS.

Number of Sides	3	4	5	6	7	8	9	10	11	12
Radius	56	70	74	60	60	98	22	89	80	85
Side	97	99	87	60	52	75	15	95	45	44

Thus, having to set out a pentagon with a side of six inches, the bevel is set to the figures in column 5, the lesser number on the tongue. In this case $74/8 = 9\frac{1}{4}$ on the tongue, and $87/8 = 10\,7/8''$ on body of the square. Sliding the bevel to 6 upon the body, the length of the radius, $5\,3/32$ will be read upon the tongue.

Problem 16.—*To divide the circumference of a circle into a given number of equal parts.*

From the column marked Y in the following table, take the number opposite the given number of parts. Multiply it by the radius of the circle, the product will be the length of chord to set off upon the circumference.

TABLE OF CHORDS OR EQUAL PARTS.

No. of Parts		Y	Z	No. of Parts	Y	Z	No. of Parts	Y	Z
3	Triangle	1.732	.5773	15	.4158	2.4050	40	.1569	6.3728
4	Square	1.414	.7071	16	.3902	2.5628	45	.1395	7.1678
5	Pentagon	1.175	.8006	17	.3675	2.7210	50	.1256	7.9618
6	Hexagon	1.000	1.0000	18	.3473	2.8793	54	.1163	8.5984
7	Heptagon	.8677	1.1520	19	.3292	3.0376	60	.1047	9.5530
8	Octagon	.7653	1.3065	20	.3129	3.1962	72	.0872	11.462
9	Nonagon	.6840	1.4619	22	.2846	3.5137	80	.0785	12.738
10	Decagon	.6180	1.6154	24	.2610	3.8307	90	.0698	14.327
11	Undecagon	.5634	1.7747	25	.2506	3.9904	100	.0628	15.923
12	Duodecagon	.5176	1.9319	27	.2322	4.3066	108	.0582	17.182
13		.4782	2.0911	30	.2090	4.7834	120	.0523	19.101
14		.4451	2.2242	36	.1743	5.7368	150	.0419	23.866

How to Use the Steel Square

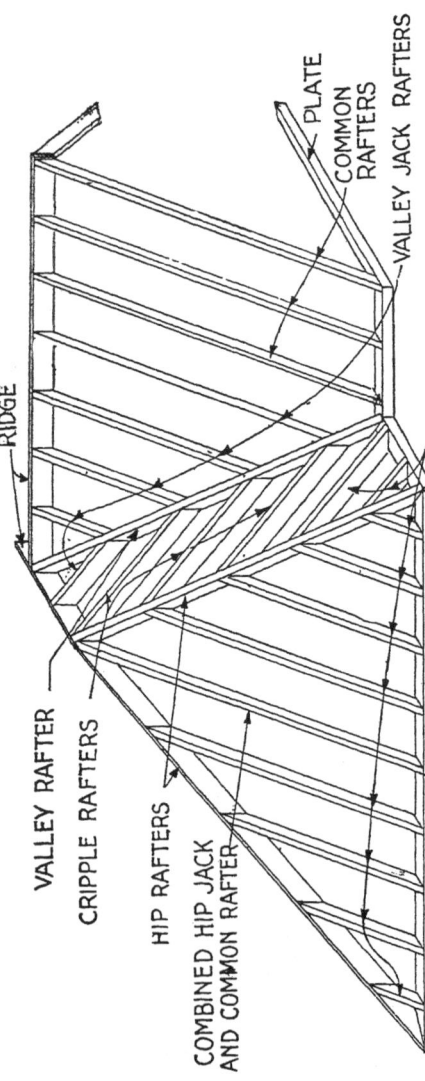

FIG. 495.—Roof frame showing ridge, plate, and different kind of rafters.

Problem 18.—Given the length of a chord, to find the radius of the circle.

This is the same as Problem 16, but the present form may be found more expeditious for calculations. The method is useful for ascertaining the diameter of gear wheels, the pitch and number of teeth having been given.

Multiply the length of the chord, width of side, or pitch of tooth by the figures found corresponding to the number of parts in column Z of the table page 343. The result is the radius of the desired circle.

Table Problems.—The term *table* is here used to denote the various markings on the framing square except the scales already described. As these tables relate mostly to problems encountered in cutting lumber for roof frame work it is necessary

How to Use the Steel Square

first to know something about roof construction so as to be familiar with the names of the various rafters and other parts.

Fig. 955 is a view of a roof frame showing the various members. In the figure it will be noticed that there is a *plate* at the

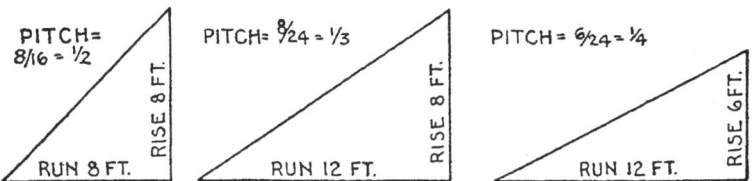

FIGS. 956 to 958.—Sections of various roofs illustrating *pitch*. To obtain the pitch: *Rule—Divide the rise by twice the run*.

bottom and ridge timber at the top, these being the main members to which the rafters are fastened.

Main or Common Rafters.—The following definitions relating to rafters should be carefully noted:

> The **rise** of a roof is *the distance found in following a plumb line from a point on the central line of the top of the ridge to the level of the top of the plate*.
>
> The **run** of a common rafter is *the shortest horizontal distance from a plumb line through center of ridge to the outer edge of the plate*.

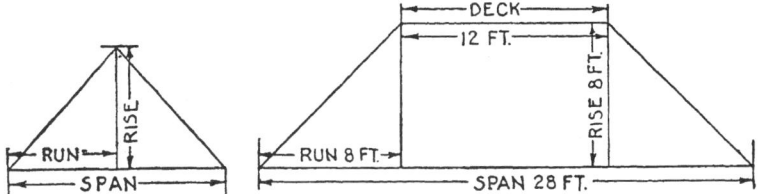

FIG. 959.—Section of roof illustrating the terms *run*, *rise*, and *span*.

FIG. 960.—Roof with *deck*. *Rule—Where rafters rise to a deck instead of a ridge, subtract the width of the deck from the span*. Here the span is 28 feet and deck 12 feet. The difference is 16 feet and the pitch is $8 \div (2 \times 8)$, or $8 \div (28-12) = \frac{1}{2}$.

The *rise per foot run is the basis on which rafter tables on some squares are made*. The term is self-defining.

To obtain the rise per foot run, multiply the rise by 12 and divide by the run, thus:

$$\text{rise per foot run} = \frac{\text{rise} \times 12}{\text{run}}$$

The factor 12 is to obtain a value in inches, the rise and run being given in feet.

Example: If the rise be 8 ft., and run 8 ft., what is the rise per foot run?
Rise per foot run $= \frac{8 \times 12}{8} = 12$ ins. The rise per foot run is always the same for a given pitch and can be readily remembered for all ordinary pitches, thus:

Pitch	½	⅓	¼	⅙
Rise per foot run (ins.)	12	8	6	4

In roof construction the rafter ends are cut to roof angles to rest respectively against ridge and plate as shown in figs. 961 and 962.

The *top* or *plumb cut* is *the cut at the rafter end which rests on the ridge*.

The *bottom* or *heel cut* is *the cut at the rafter end which rests against the plate*.

The *length* of a common rafter is *the shortest distance between the outer edge of the plate and the central line of the top of the ridge*. It should be distinctly understood that this is not the *real* length but the *artificial* length, or value, which must be used in applying the table markings on the square. The real length is obtained by subtracting half the thickness of the ridge from the artificial length.

The *pitch* is the *proportion that the rise bears to the whole width of the building (or the span).**

The pitch expressed as an equation is:

$$\text{pitch} = \frac{\text{rise}}{\text{span}} \quad \dots\dots\dots\dots\dots\dots\dots\dots\dots (1)$$

---

*NOTE.—Where rafters rise to a deck instead of a ridge, it is necessary to subtract the width of the deck from the total span.

How to Use the Steel Square

Example.—A building 24 ft. wide has a roof with a rise of 8 ft. What is the pitch of the roof?

Substituting in (1)

$$\text{pitch} = 8/24 = 1/3$$

The question is often asked, what constitutes full pitch? From an inspection of equation (1) this is easily answered. Since the pitch is full when the value of rise ÷ span = 1 then from the equation evidently the pitch is full when the rise is equal to the span that is, equal to twice the run. Accordingly

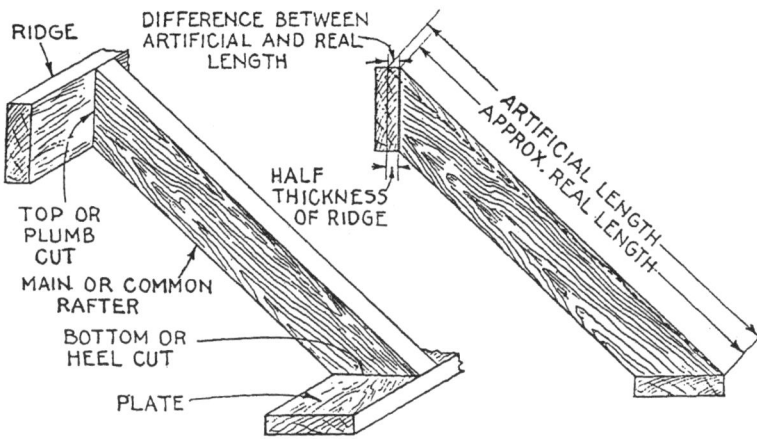

FIG. 961.—Portion of roof frame illustrating *top* or *plumb cut* and *bottom* or *heel cut*.

FIG. 962.—End view of portion of roof frame illustrating artificial and real length of rafter. Rafter-tables as given on framing squares are figured for an arbitrary length being the shortest distance from the outer edge of the plate to the center of the top of the rafter. This would be the actual length of the rafters if they rested at the upper end against each other instead of against the ridge. This arbitrary or artificial length must be assumed as a basis for the rafter table, otherwise separate values would be necessary for various thicknesses of the ridge member and it would be not only confusing but impossible to put all the figures in the limited space available on the square. Hence, to obtain approximate real length of rafter *subtract half thickness of ridge from the artificial length or value given on the square*. It should be understood that this is the ***approximate real length***, or near enough for practical use. However, the enlightened carpenter will want to know what is the ***actual real length*** and why it is not used in practice as explained in fig. 963. Note in fig. 963, that the rafter, as cut, is too short.

for full pitch if the run be say 12 ft., the rise is 24 ft. With this as a basis a table of various pitches made thus:

Pitch Table

Pitch	1	$^{11}/_{12}$	$^{5}/_{6}$	$^{3}/_{4}$	$^{2}/_{3}$	$^{7}/_{12}$	$^{1}/_{2}$	$^{5}/_{12}$	$^{1}/_{3}$	$^{1}/_{4}$	$^{1}/_{6}$	$^{1}/_{12}$
Run	12	12	12	12	12	12	12	12	12	12	12	12
Rise	24	22	20	18	16	14	12	10	8	6	4	2

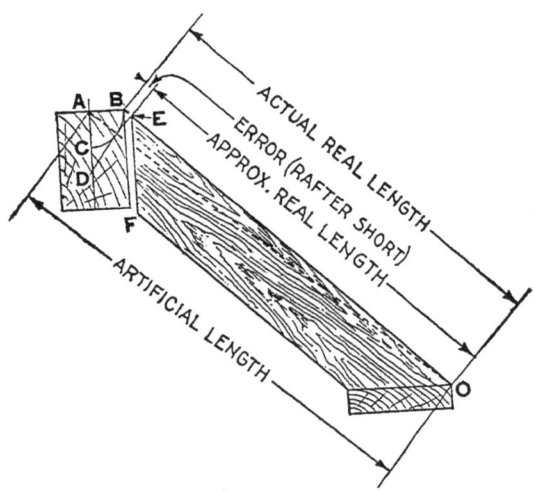

FIG. 963.—Detail of rafter and ridge illustrating why rafters are cut only to approximate real length. OA, is the length the rafter would be if there were no ridge board; this is the length found on the square. *In cutting* a rafter to fit against a ridge, an allowance must be made for the space taken up by the ridge each side of the center line, as AB. Hence OB, is the actual or real length of the rafter, *but this length cannot be conveniently found. In practice* therefore, an approximation is made by subtracting from the artificial real length OA, or value given on the square, an amount AB, equal to half the thickness of the ridge board. With A, as center and AB, as radius, describe arc BC, and with O, as center describe arc DE, tangent to BC. This gives the point E, such that OE, is the approximate rafter length, that is, the length OA, less half thickness of the ridge, the upper end of the rafter being indicated by the line EF, the rafter being actually too short by the distance BE. In the diagram, BE, appears large because a very thick ridge was selected to augment the error. *In practice* with a ridge of normal thickness, the error BE, is very small and when the rafter is in place is hardly noticeable.

How to Use the Steel Square

Now with a 24 in. square diagonals connecting 12 on the tongue (corresponding to the run) and value from table for rise will give pitch angle for any combination or run and rise.

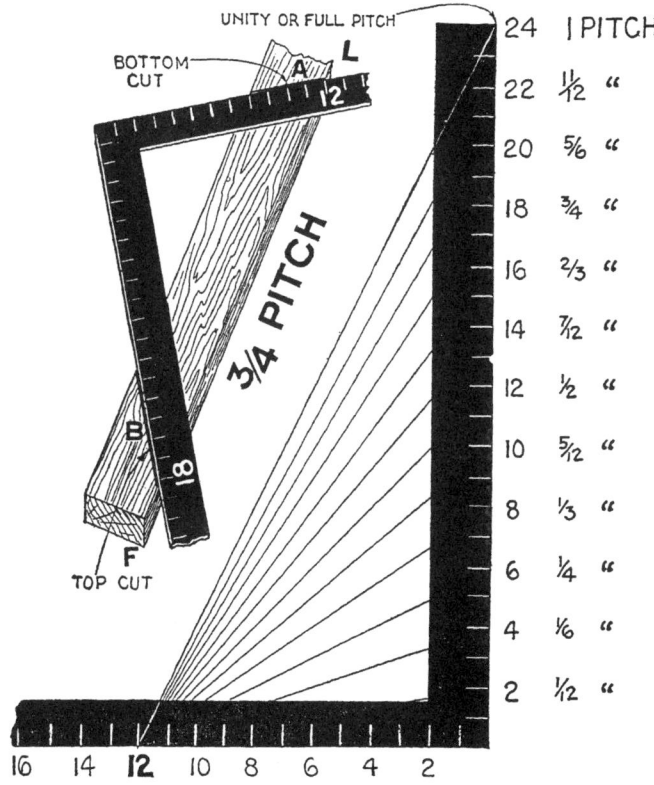

FIG. 964 and 965.—Application of the framing square for obtaining various pitches as given in the accompanying table. In fig. 964 the square is seen applied to a rafter with the 12-inch mark on tongue and 18-inch mark on body at the edge of the rafter. The inclinations A, and B, of the tongue and body of the square with the edge LF, of the rafter give the correct angles for cutting so that edge F, will have ¾ pitch when placed in position, that is, when A, is horizontal and B, vertical or plumb.

22   *How to Use the Steel Square*

Thus lay off 12 on tongue and 8 on body for ⅓ pitch. The various pitches given in the table are shown in fig. 965.

Hip (and Valley) Rafters.—The hip rafter represents *the hypothenuse or diagonal of a right-angle triangle*, one side being the *common rafter*, and the other side the *plate*, or that part of the plate lying between the foot of the hip rafter and the foot of the adjoining common rafter as shown in fig. 966.

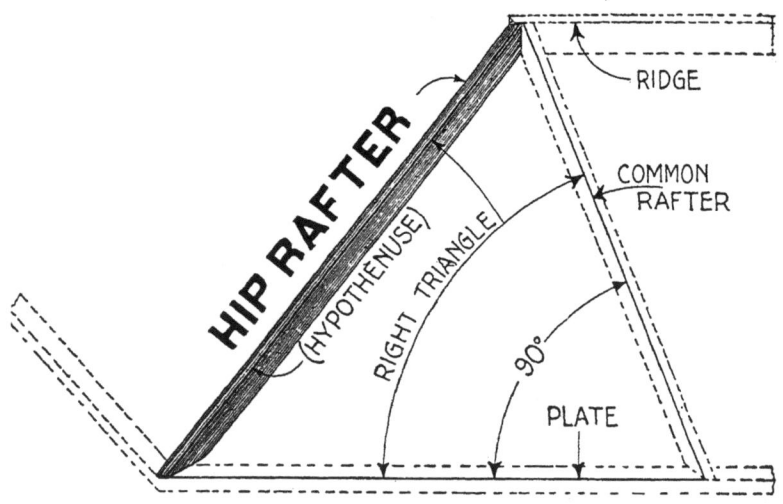

FIG. 966.—Hip rafter as framed between plate and ridge, showing that the hip rafter is *the hypothenuse of a right angle triangle*, whose other two sides are the adjacent common rafter and intercepted portion of the plate.

The rise of hip rafter is the same as common rafter. The run of the hip rafter is the horizontal distance from the plumb line of its rise to the outside of the plate at the foot of the hip rafter. This run of the hip rafter is to the run of the common rafter as 17 is to 12. Therefore, for ⅙ pitch the common rafter run and rise are 12 and 4, while the hip rafter run and rise are 17 and 4.

For the top and bottom cuts of the common rafter, the figures are used that represent the common rafter run and rise, that is, 12 and 4 for ⅙ pitch, and 12 and 6 for ¼ pitch, etc., but for top and bottom cuts of hip rafter use the figures 17 and 4, and 17 and 6, etc., the run and rise of the hip rafter.

Valley Rafters.—The valley rafter is the hypotenuse of a

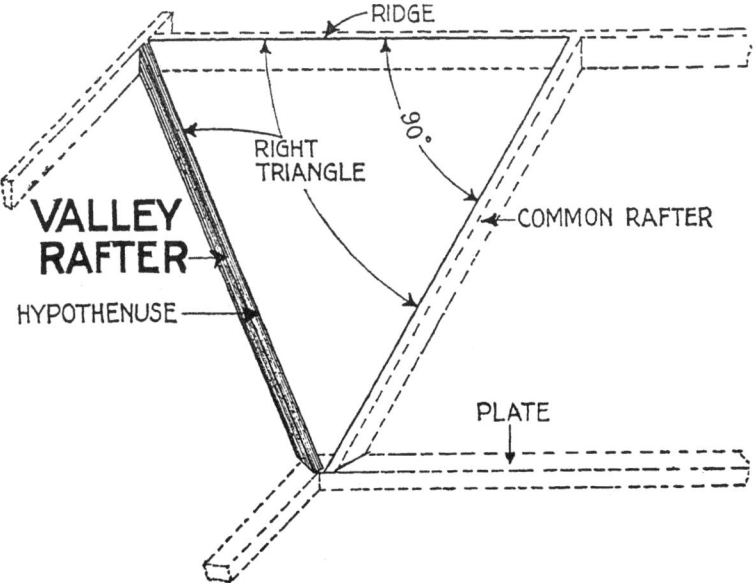

FIG. 967.—Valley rafter as framed between plate and ridge, showing that the valley rafter is the hypotenuse of a right triangle whose other two sides are the adjacent common rafter and intercepted portion of the ridge.

right angle triangle made by the common rafter with the ridge, corresponding with the right angle triangle made by the hip rafter with common rafter and plate; and, therefore, the rule for the lengths and cuts of valley rafters are the same as for hip rafter.

Jack Rafters.—These are usually spaced either 16 inches apart or 24 inches apart, and, as they lie against the hip or valley equally spaced, the second jack rafter must be twice as long as the first, the third three times as long as the first, and so on. The reason for 16 and 24 inch spacing on jack rafters is because laths are 48 inches long, therefore the rafter must be 16 or 24 inches so the lath may be nailed to it.

Cripple Rafters.—A cripple rafter is *a jack rafter which touches neither the plate nor the ridge; it extends from valley rafter to hip rafters.*

Cripple rafter length is that of the jack rafter plus length necessary for its bottom cut, which is a plumb cut like top cut. Top and bottom (plumb) cuts of cripples are same as top cut for jack rafter. Side cut at hip and valley same as side cut for jacks.

12, 13, 17

Finding Rafter Lengths Without Aid of Tables.—In the directions accompanying framing squares and in some books frequent mention is made of the figures **12, 13,** and **17,** thus for common rafters "use figure **12** on body and rise of roof on tongue;" for hip or valley rafters, "use figure **17** on body and rise of roof on tongue"—and no explanation of how these fixed numbers are obtained. The intelligent workman should not be satisfied with knowing which number to use but he should

---

NOTE.—In the tables the common rafter length is given to the center of the ridge, and so with the hip rafter length. The jack rafter length is given to the center of its hip or valley rafter. In using the tables, make allowances indicated, depending upon the thickness of ridge, hip and valley rafters.

How to Use the Steel Square

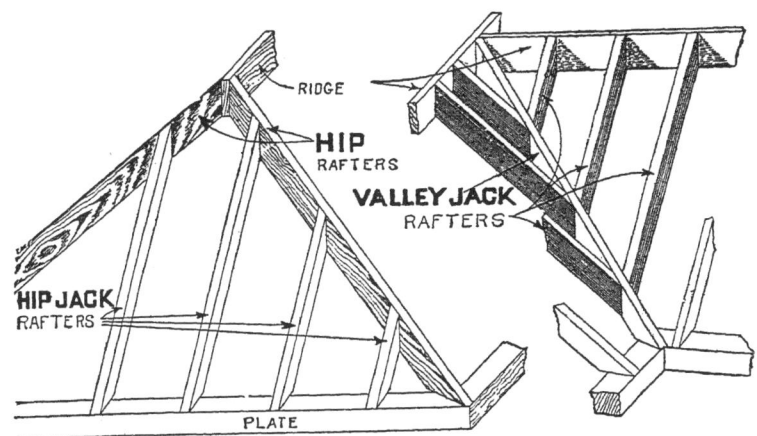

FIG. 968.—Hip jack rafters as framed *between the plate and hip rafters*.

FIG. 969.—Valley jack rafters as framed between the valley rafter and ridge.

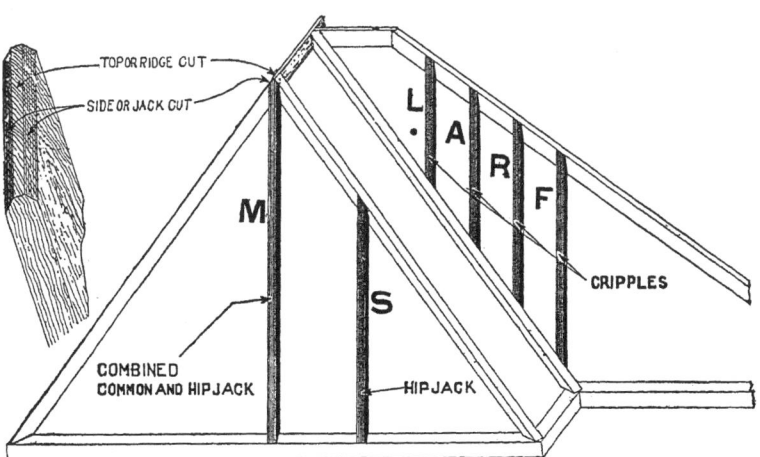

FIG. 970.—Combined common and hip jack rafter and cripple rafters. When a jack as rafter **M,** joins with the end of the ridge it forms a combination of common and hip jack rafters, because at the ridge end it has the common rafter plumb cut and the hip jack side cuts, these being plainly shown in the enlarged detail of end. **S,** is a hip jack rafter. **L, A, R, F,** are cupple rafters.

want to know *why* each particular number is used. This is easily understood by aid of fig. 971.

Here let ABCD be a square whose sides are 24 ins. long, and *abcdefg* L, an inscribed octagon. Each side of the octagon as *ab*, *bc*, etc., will measure 10 ins., that is, LF = one-half side = 5 ins. and by construction FM, = 12 ins. Now let FM, represent the run of a *common* rafter. Then LM, will be run of an *octagon* rafter, and DM, run of a hip or valley rafter. The

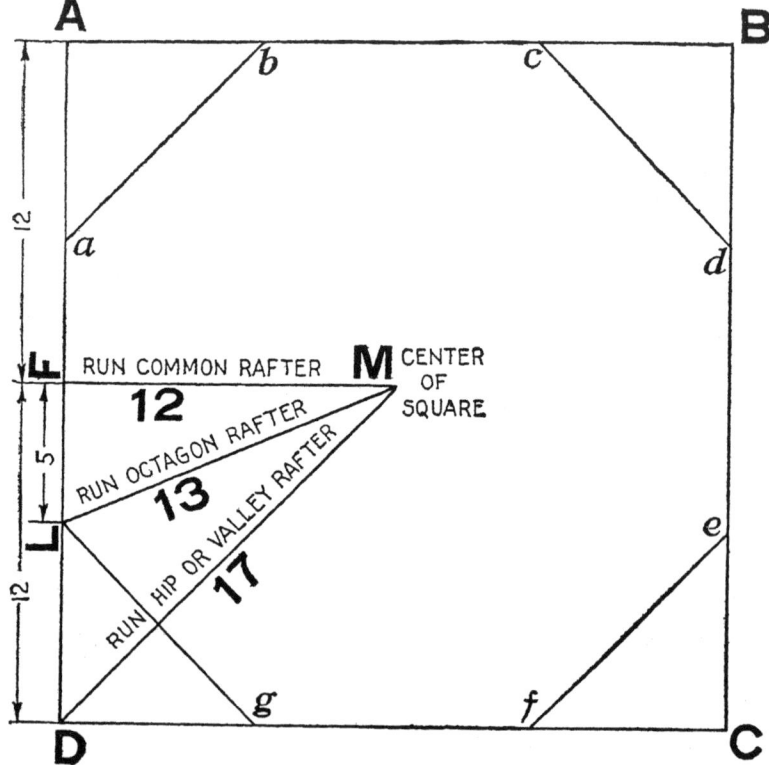

FIG. 971.—Square and inscribed octagon illustrating method of obtaining and use of points 12, 13, and 17 in application of square in obtaining length of rafters without aid of rafter tables.

How to Use the Steel Square

values for run of octagon and hip or valley rafters (LM and DM) are obtained thus:

$$LM = \sqrt{FM^2 + LF^2} = \sqrt{12^2 + 5^2} = 13$$
$$DM = \sqrt{FM^2 + DF^2} = \sqrt{12^2 + 12^2} = 16.97, \text{ say } 17$$

Example.—What is the length of a *common* rafter having a 10 foot run and ⅜ pitch?

For a 10 ft. run the span = 2 × 10 = 20 ft. and with ⅜ pitch
rise = ⅜ of 20 = 7.5 ft.

$$\text{rise per foot run} = \frac{\text{rise} \times 12}{\text{run}} = \frac{7.5 \times 12}{10} = 9 \text{ ins.}$$

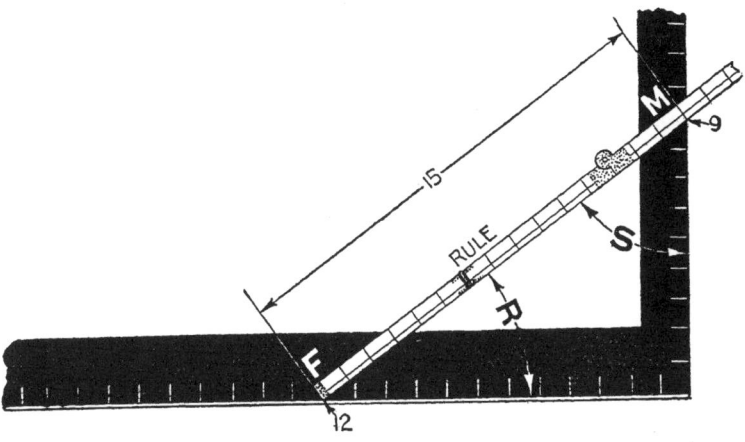

FIG. 972.—Rule placed on square at points 12 and 9, for obtaining length of common rafter per foot run with ⅜ pitch.

On the body of the square, fig. 972, take 12 ins. for 1 ft. of run, and on the tongue, 9 for rise per foot run. The diagonal or distance between the points thus obtained will be length of common rafter per foot run with ⅜ pitch. This distance FM, in fig. 971 measures 15 ins., or by calculation:

$$FM = \sqrt{12^2 + 9^2} = 15 \text{ ins.}$$

Since the length of run is 10 ft.,
length of rafter = 10 × length per foot = 10 × 15 = 150 ins., = 150 ÷ 12 = 12½ ft.

The combination of figures 12 and 9 on the square as in fig. 972 not only gives the length of rafter per foot run, but if the rule be considered as the rafter, the angles S and R, for top and bottom cuts are obtained. The points for making are accordingly found *by placing the square upon the rafter so that a portion of one arm of the square represents the run and a portion of the other arm, the rise.* For the common rafter with ⅜ pitch the points are 12 and 9, the square being placed on the rafter as in fig. 973.

Example.—What is the length of an *octagon* rafter to join a common rafter having a 10 ft. run (as rafters MF and ML in fig. 971)?

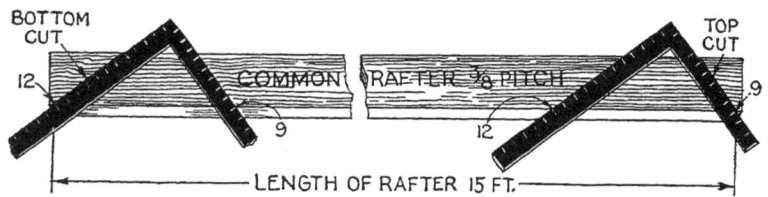

FIG. 973.—Square placed on rafter at points 12 and 9, which give the proper angles for bottom and top cuts.

From fig. 971 it is seen that the run per foot of an octagon rafter as compared with a common rafter is as 13 to 12, and that the rise per 13 ins. run of octagon rafter is the same as per 12 ins. of common rafter. Hence measure across from point 13 and 9 on the square as MS in fig. 974, which gives length (15¾ ins.) of octagon rafter per foot run of common rafter. This length multiplied by run of common rafter gives length of octagon rafter, thus:

$$15¾ \times 10 = 157½ \text{ ins.} = 13 \text{ ft. } 1½ \text{ ins.}$$

Points 13 and 9 on the square (MS in fig. 974) give angles for top and bottom cuts.

Example.—What is the length of a hip or valley rafter to join a common rafter having a 10 ft. run (as rafters MF and MD in fig. 971)?

Fig. 971 shows that the run per foot of a hip or valley rafter as compared with a common rafter is as 17 to 12 and that the rise per 17 ins. run of hip or valley rafter is the same as per 12 ins. of common rafter. Hence measure across from points 17 and 9 on the square as LS in fig. 974 which gives length

How to Use the Steel Square

(19¼ ins.) of hip or valley rafter per foot of common rafter. This length multiplied by run of common rafter gives length of hip or valley rafter, thus:

$$19\tfrac{1}{4} \times 10 = 192\tfrac{1}{2} \text{ ins.} = 16 \text{ ft. } \tfrac{1}{2} \text{ in.}$$

Points 17 and 9 on the square (LS in fig. 974) give the angles for top and bottom cuts.

The following table gives the points on square for top and bottom cuts of various rafters.

Square Points for Top and Bottom Cuts

PITCH		1	11/12	5/6	3/4	2/3	7/12	1/2	5/12	1/3	1/4	1/6	1/12
Tongue	Common							**12**					
	Octagon							**13**					
	Hip or Valley							**17**					
Body		24	22	20	18	16	14	12	10	8	6	4	2

Rafter Tables.—The arrangement of these tables varies considerably with different makes of square, not only in the way it is calculated but also in its position on the square. On some squares the rafter table is found on the face of the body, on others, on the back of the body. There are two general classes of rafter table, grouped according as the figures give:

1. Total length of rafter, or
2. Length of rafter per foot run.

Evidently where the total length is given there is no figuring to be done, but when the length is given per foot run, the reading must be multiplied by the length of run to obtain total length of rafter. To illustrate these differences directions for using several well known squares will now be given. These differences relate to the common and hip or valley rafter tables.

Class 1.—*Reading total length of rafter*

The Sargent square is selected as an example of **Class 1** reading rafter lengths direct without any figuring. The rafter tables occupy both sides of the body instead of being combined in one

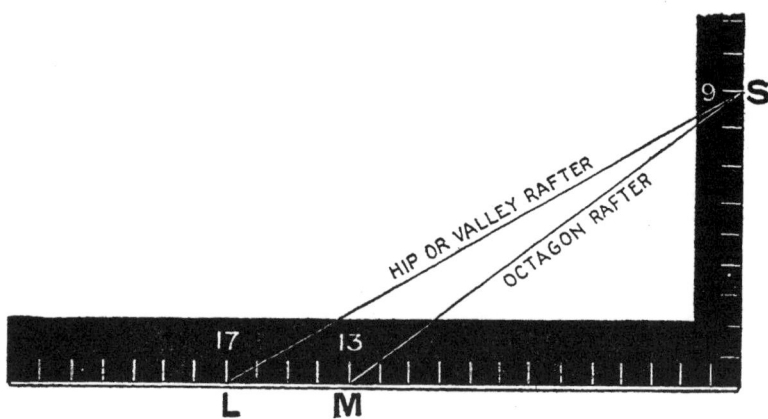

FIG. 974.—Measurements across square for octagon, and hip or valley rafters illustrating use of points **13** and **17**. MS, (**13,** 9) octagon rafter length per foot run of common rafter; LS, (**17,** 9) hip or valley rafter per foot run of common rafter ⅝ pitch.

table; that is, the common rafter table is found on the back side *and the hip, valley and jack* table on the face side.

Common rafter table.

The common rafter table, fig. 975, includes the outside edge graduations of the back of the square on both body and tongue and is in twelfths. The inch marks may represent inches or feet, and the twelfth marks, twelfths of an inch or twelfths of a

How to Use the Steel Square

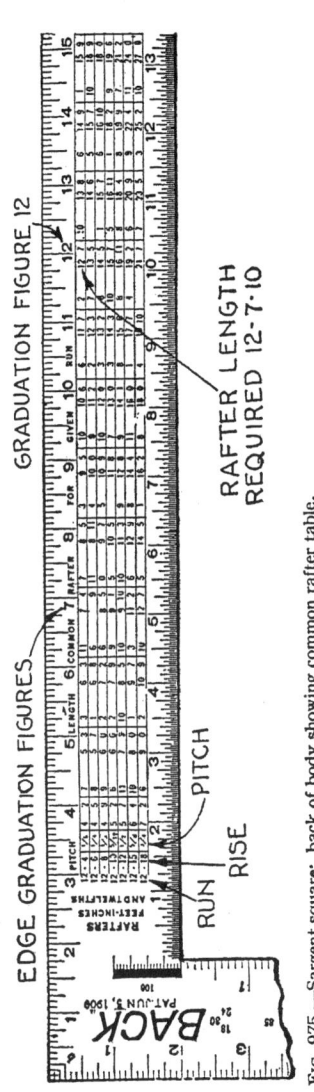

FIG. 975.—Sargent square; back of body showing common rafter table.

foot (that is inches), as a scale. The *edge graduation figures* above the table represent the *run* of the rafter and under the proper figure on the line representing the *pitch* will be found in the table, the rafter length required. The pitch is represented by the figures at the left of the table under the word PITCH thus:

12 feet run to 4 feet rise is 1/6 pitch
12   "   "   6   "   "   " 1/4   "
12   "   "   8   "   "   " 1/3   "
12   "   " 10   "   "   " 5/12 "
12   "   " 12   "   "   " 1/2   "
12   "   " 15   "   "   " 5/8   "
12   "   " 18   "   "   " 3/4   "

The length of common rafter given in table is *from top center of ridge board to outer edge of plate*. In actual practice deduct for one-half thickness of ridge board and add for any projection of eave beyond the plate.

To Find the Length of a Common Rafter.—For a roof with ⅙ pitch (that is, rise = ⅙ the width of the building) and having a run of 12 feet, follow in the common rafter table (fig. 975) the upper or ⅙ pitch ruling.

Find under the graduation figure 12, the rafter length required which is 12, 7, 10 which means 12 feet $7^{10}/_{12}$ ins. If the run be 11 feet and the pitch ½ (or the rise ½ the width of the building) then the rafter length will be 15, 6, 8, which means 15 feet 6 $^8/_{12}$ ins.

Again, if the run be 25 ft., add the rafter length for run of 20 ft. to the rafter length for run of 5 ft. When the run is in inches, then in the rafter table read inches and twelfths instead of feet and inches. For instance, if with ½ pitch the run be 12 ft. 4 ins., add the rafter length of 4 ins. to that of 12 ft. as follows:

For run of 12 ft. the rafter length is 16 ft. 11 $^8/_{12}$ ins.

For run of 4 ins. the rafter length is   5 $^8/_{12}$ ins.

Total 17 ft. 5 $^4/_{12}$ ins.

The run of 4 ins. is found under the graduation 4 and is 5, 7, 11, which is approximately 5 $^8/_{12}$ ins. If it were ft. it would read 5 ft. 7 $^{11}/_{12}$ ins.

Hip rafter table.

This table as shown in fig. 976 is on the face of the body and is used substantially as the table for common rafters just explained. In connection with the hip rafter table the outside edge graduation figures represent the *run* of common rafters. The length of rafter given in the table is from *top center of ridge board to outer edge of plate*. In actual practice, deduct ½ thickness of ridge board and add for any projection beyond the plate for eave. In using the table, seek the figures on the line with the required pitch of the roof.

Under heading "Pitch" the set of three columns of figures gives the *pitch;* the seven pitches in common use, as ⅙ — 12 — 4 (for each 12″ run a 4″ rise).

How to Use the Steel Square

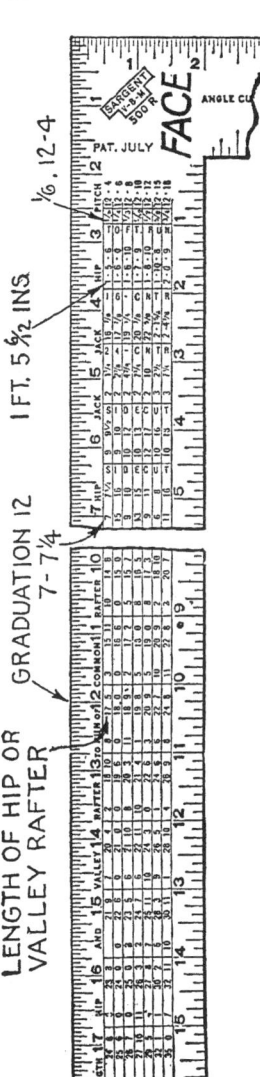

FIG. 976.—Sargent square; face of body showing hip rafter table and hip or valley readings referred to in the text.

Under heading "Hip" the set of figures gives for each pitch the length of hip and valley rafter per foot of run of common rafter, as 1 ft. 5 6/12 ins. for 1/6 pitch.

Under heading "Jack" (16 ins. on center) the set of figures gives the length of the shortest jack rafter, spaced 16 ins. on center which is also the difference in length of succeeding jack rafters.

Example.—If the jack rafters be spaced 16 ins. on center for a 1/6 pitch roof, find lengths of jacks and cut bevels.

NOTE.—Home made *fence* for square. A *fence* may be defined in general as *a guide fitted to a tool*. As adapted to the stee square it is a long strip arranged to be clamped to the body and tongue in any angular position, being useful where a number of pieces such as rafters must be similarly marked. A good home made fence for the square may readily be made from a stick of hard wood about 2 ins. wide, 1½ ins. thick and 2½ ft. long. Cut a saw kerf from both ends into which the saw will slide, leaving about 8 ins. solid wood near the middle. The fence is secured to the square in any desired position by means of clamp bolts. A series of holes may be bored through the fence to receive the bolts or in place of these narrow slots may be cut each side of the solid middle section, extending almost to the ends of the fence, and in which the bolts may slide to desired position and there clamped. Evidently where a number of pieces are to be similarly marked (as rafters) they may be uniformly marked by using a fence, whereas with independent setting each time no two would be marked exactly alike, owing to the difficulty of placing the setting numbers exactly over the edge of the timber, especially when hurriedly done.

How to Use the Steel Square

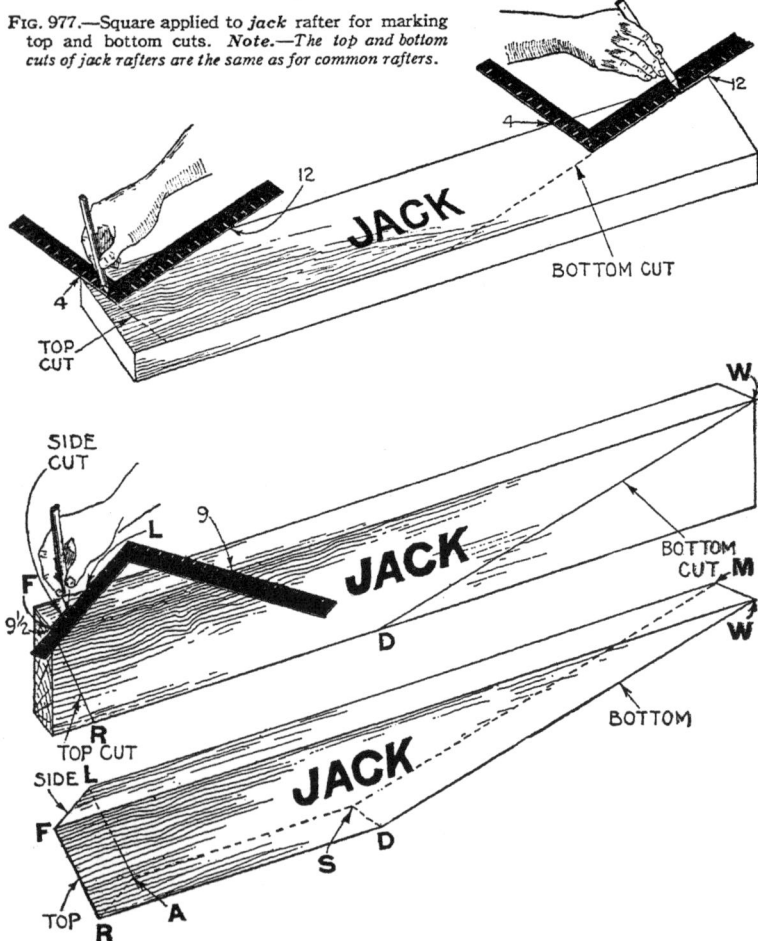

FIG. 977.—Square applied to *jack* rafter for marking top and bottom cuts. *Note.—The top and bottom cuts of jack rafters are the same as for common rafters.*

FIG. 978.—Square applied to *jack* rafter for marking side cut. Here FR, and DW, are the marks for top and bottom cuts previously marked in fig. 977.

FIG. 979.—*Jack* rafter cut as marked in fig. 978 LARF, shows section cut at top end of rafter and MSDW, section cut at bottom end.

How to Use the Steel Square

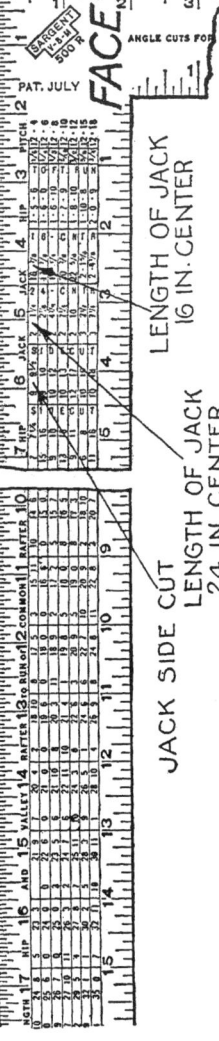

FIG. 980.—Sargent square; face of body showing rafter table and jack readings referred to in the text.

The jack top and bottom cuts (or plumb and heel cuts) are same as for common rafter. Take 12 on tongue of square (as before explained) and on the body take the figure which represents the rise per foot of the roof, or if the pitch be given take the figures in the table on page 357 corresponding to the given pitch. Thus for $\frac{1}{6}$ pitch these points are 12, 4.

Fig. 977 shows square on jack in this position for marking top and bottom cuts.

Look along the line of $\frac{1}{6}$ pitch (fig. 980) under *jack* (16 in. center) and find $16\frac{7}{8}$ which is the length in *inches* of the shortest jack and is also the amount to be added for the second jack. Deduct for half the thickness of hip rafter because jack rafter lengths given in table are to centers. Also add for projection beyond outer edge of plate if any.

Look along the line of $\frac{1}{6}$ pitch (fig. 980) under **Jack** (*side cut*) and find $9\text{-}9\frac{1}{2}$ for $\frac{1}{6}$ pitch. These figures refer to the graduated scale on the edge of the arm of the square.

To obtain the required bevel take 9 on one arm and $9\frac{1}{2}$ on the other as shown in fig. 978.*

It should be carefully noted that the **last figure or figure to the right** gives the point on the marking side of the square, that is, mark on the $9\frac{1}{2}$ side as shown in the figure.

---

*NOTE.—The figures for side cuts of jacks are also correct for cuts of the valley moulding at the junction of two gables, etc. In stair building as a rule the stringers rise on a rough floor, hence, allow for the thickness of finished floor in measuring for the first rise.

Under heading "Jack" (*24 ins. on center*) the set of figures gives the length of the shortest jack rafter spaced 24 ins. on center, which is also the difference in length of succeeding jack rafters, as 2 ft. 1¼ ins. for ⅙ pitch. Deduct for half the thickness of hip or valley rafter because jack rafter lengths given in the table are to centers. Also add for projection beyond the plate.

Under heading "Hip" the set of figures gives the side cut of hip and valley rafter against ridge board or deck, as 7—7¼ for ⅙ pitch (mark on the 7¼ side).

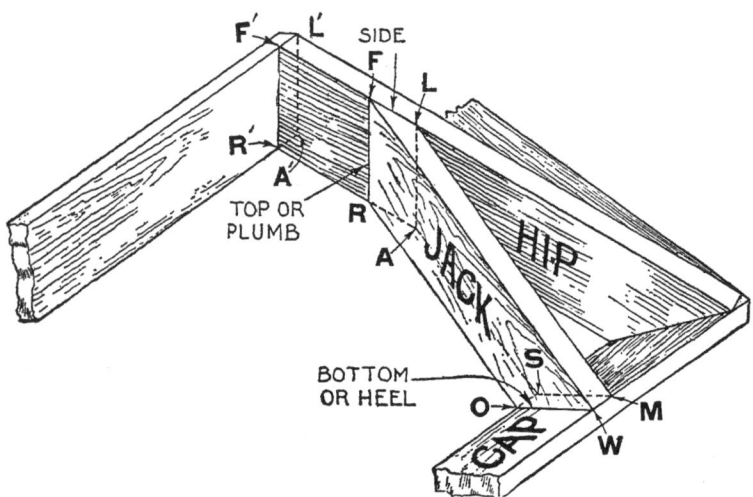

FIG. 981.—*Jack* rafter (as cut by method in figs. 977 to 979) in position on roof showing fit at ends with hip rafter and cap. Here as in fig. 979 LARF, shows section cut at top end of jack, and MSODW, section cut at bottom end.

To get the cut of the sheathing and shingles (whether hip or valley) reverse the figures under hip—as 7¼—7 instead of 7—7¼.

For (hip) top and bottom cuts take 17 on body of square and take on tongue the figure which represents the rise per foot of the roof.

Figs. 982 to 984 show marking and cut of hip rafter and fig. 981 rafter in position resting on cap and ridge. Here the section L′A′R′F′ resting on ridge is the same as L′A′R′F′ in fig. 984.

Under heading "Hip and Valley" the set of figures gives for each pitch

How to Use the Steel Square

the length of hip or valley rafter of run of common rafter. For instance, for roof having ⅙ pitch under the figure 12 (representing the run of common rafter or half the width of the building) along the ⅙ pitch line of figures find 17, 5, 3 which means 17 ft. 5 3/12 ins., length of hip or valley rafter. Deduct for half the thickness of the ridge board and add for eave overhang beyond the plate, which is the length of hip or valley rafter required for roof of ⅙ pitch and common rafter run of 12 ft.

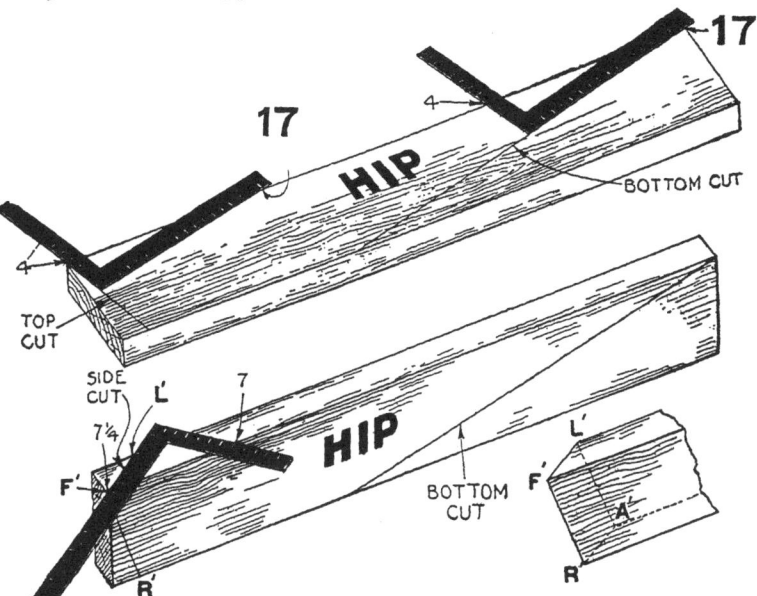

FIG. 982.—Square applied to *hip* rafter for marking top and bottom cuts. Note that the fixed number *17* is used for hip rafters.

FIG. 983.—Square applied to *hip* rafter for marking side cut. Here F'L' is the mark for side cut and F'R' the mark for top cut previously made in fig. 982.

FIG. 984.—Detail of upper end of *hip* rafter showing section L'A'R'F, cut to bevel required for ridge.

Example.—Find the length of hip rafter for a building 24 ft. span, ⅙ pitch (4 ins. rise per ft.).

In the hip rafter table (fig. 976) along the line of figures for ⅙ pitch and under the graduation figure 12 (representing half the span or run of common rafter) find 17, 5, 3 which means 17 ft. 5 $^3/_{12}$ ins. the required length of hip or valley rafter. Deduct for half the thickness of the ridge board and add for any overhang required beyond the plate.

For top and bottom cuts of hip or valley rafter take 17 on body and 4 (the rise of the roof per ft.). Mark on 17 side gives the bottom cut, on 4 side, the top cut.

For *side cut* of hip or valley rafter against ridge board look in the set of figures for *side cut* in the table (fig. 976) under *hip*, along the line for ⅙ pitch and find the figure 7, 7¼. Use 7 on one arm of square and 7¼ on the other; mark on the 7¼ arm for side cut.

Class II.—*Reading length of rafter per foot run*

There are numerous squares having rafter tables based on "run per foot," such as the Southington Hardware Co., Stanley, Eagle, etc. Of these, the arrangement of the rafter tables are identical on the Stanley and Eagle squares, but considerably different on the Southington Hardware Co. square.

Fig. 985 shows the rafter table of the Southington Hardware Co. square. As will be seen there are several combinations of figures corresponding to headings: "Length common rafters per foot run," "Length of hip or valley rafters per foot run," etc., the values being found to the right of each heading. To demonstrate how to use these values or figures take the first combination of figures headed "Length of common rafter per foot run." To the right of this heading are several pairs of figures, the upper figure of each pair, as 3, 4, 6, 8, 10, 12, 15, 16, 18, means *rise (in ins.) per foot run*. Under each of these numbers is a number which means the *length of rafter per foot run*, corresponding to the rise given immediately above.

Example.—Find the length of a common rafter on a building 24 ft. wide and ½ pitch.

How to Use the Steel Square

The rise per foot run = pitch × span ÷ run

= ½ × 24 ÷ 12 = 1 ft. or 12 ins.

In the combination of numbers headed "Length of common rafter per foot run" (fig. 985) look for 12, and under the number will be found 16.97, which is the length of a common rafter per foot run. Hence, since run is 12 ft.

length common rafter = 16.97 × 12 = 203.64 ins., or

203.64 ÷ 12 = 16.97 ft.

This reading is shown in detail in fig. 986.

The other combinations of figures for hip or valley rafters, jacks are read in a similar manner rendering further explanation unnecessary.

The Eagle rafter table, as shown in fig. 987, is located on the face of the body of the square and is composed of six rows of figures which are lettered at the left end of the body to show their use.

The figures found in these six rows refer to the outside edge graduations which in the case of the side cuts are clearly marked beyond all mistake.

The inch marks may represent inches or feet, and the twelfths may represent twelfths of an inch or twelfths of a foot that is regarded as a

FIG. 985.—Southington Hardware Co. square; back of tongue showing rafter table and illustrating *Class II* in which the reading gives *length of rafter per foot run*, and the grouping of rafter values, the latter feature being special with this square.

scale. The edge marks represent the rise of a roof as 4 inches to the foot run called ⅙ pitch, or 6 inches to the foot run called ¼ pitch. After looking at the inch line figures on the outside and finding the figure that is the same as the rise of the roof, look underneath it and find a table giving the lengths of rafters and all side cuts. The run in every table is 1 foot. There are seventeen of these tables commencing at two inches and continuing to eighteen inches.

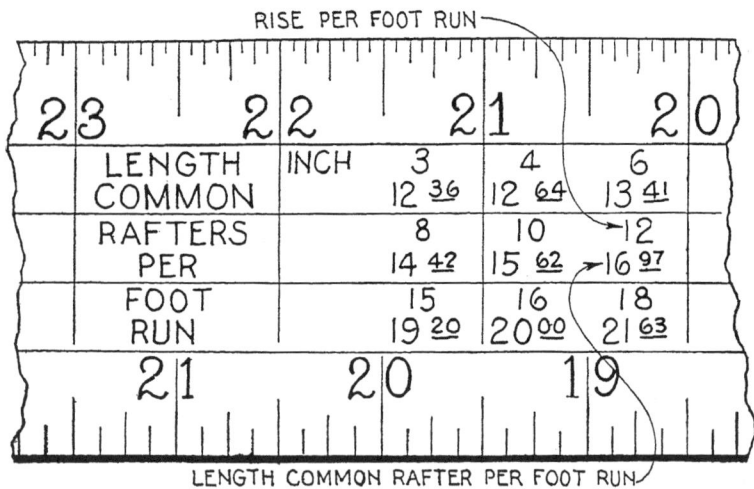

FIG. 986.—Southington Hardware Co. square; enlarged portion of rafter table showing reading referred to in the text.

Example.—Find the length of a common rafter for a roof having 8 in. rise or ⅓ pitch. 20 ft. span under 8 on the upper edge scale of the square, fig. 987, will be found a table and the first figures of the table which is designated at the left end of the body, *"Length of main rafters per foot run"* are 14.42. Multiply this by half of the width of the building, which will give the whole length of the rafters, thus:

$$14.42 \times \tfrac{1}{2} \text{ of } 20 = 144.2 \text{ ins., or}$$
$$144.2 \div 12 = 12.02 \text{ ft.}$$

How to Use the Steel Square

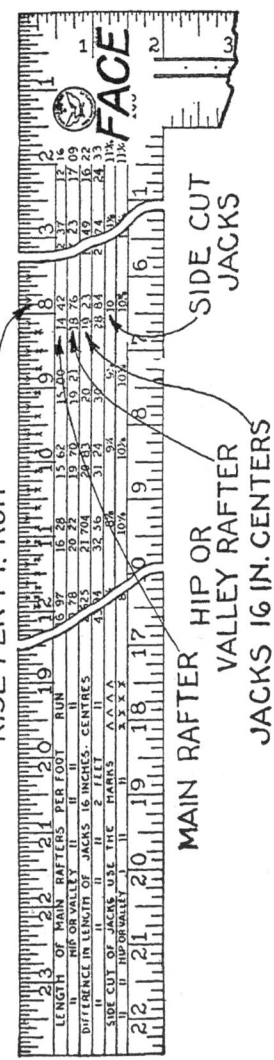

FIG. 987.—Eagle square; face of body showing rafter tables and readings referred to in the text.

Example.—Find the length of a hip or valley rafter for a roof having 8 in. rise per foot run or ⅓ pitch 20 ft. span. On the outer edge scale, fig. 987, find 8. Look below this number on second line marked *"Length of hip or valley rafters per foot run"* and find 18.76, then

Length of hip or valley rafter = 18.76 × ½ of 20 = 187.6 ins. = 187.6 ÷ 12 = 15.63 ft.

Example.—Find the lengths of jack rafters for a roof with 8 in. rise per foot run ⅓ pitch 16 in. centers; 2 ft. centers.

In fig. 987 take 8 on the outer edge scale to represent the rise. Look under the figure in the third line marked *"Difference in lengths of jacks 16 in. centers"* and find 19.23 ins. This is the length of the first jack rafter when they are spaced 16 ins. between centers, and it is also the difference between the lengths of the others, each one being 19.23 ins. longer than the one nearer the first one. The figure immediately below in the fourth line, 28.84 ins. is the length of the first jack when they are spaced 24 ins. between centers and is also the difference in lengths of the others.

Example.—Find side cuts on jacks and hip and valley rafters for roof with 8 in. rise or ⅓ pitch.

The numbers 8 and 12 are points for top and bottom cuts as before explained. In fig. 987, take 8 on the upper edge scale to represent the rise Look under the figure in the fifth line marked "*Side cut of jacks use the marks* ∧ ∧ ∧ ∧" and id 10. This refers to the graduation marks on the outside edge of the body.

Set square on jack to these marks and mark along the 12 side for cut of jack. This also gives the right angle to cut plancier and moulding on the jet that runs up the gable.

The level plancier and moulding cuts can be marked on the body side or the references transposed using the 12 in. mark on body and reading given in the table on the tongue.

Side cuts for hip and valley rafters are found by using the figures in the bottom line in the same way as just explained for jacks.

It should be noted that the 12 in. mark on the tongue is always used in all angle cuts, both top and bottom and side cuts, thus leaving the workman but one number to remember when laying outside or angle cuts. This is the figure taken from the fifth or sixth number in the table. The side cuts come always on the right hand or tongue side on rafters. When marking boards these can be reversed for convenience at any time by taking the 12 in. mark on the body and using the references on the tongue.

Table of Octagon Rafters.—The Eagle square is provided with a table for cutting octagon rafters as shown in fig. 988.

FIG. 988.—Eagle square; back of body showing octagon rafter tables and reading referred to in the text.

How to Use the Steel Square

In this table the first line of figures from the top gives the length of *octagon hip rafters* per foot of run.

The second line of figures gives length of *jack rafter* for one foot space from octagon hip.

The third line of figures gives the reference to the graduated edge that will give the side cut for octagon *hip rafters*.

The fourth line of figures gives the reference to the graduated edge that will give the side cuts for *jacks*.

The tables are used in a manner similar to that used for the

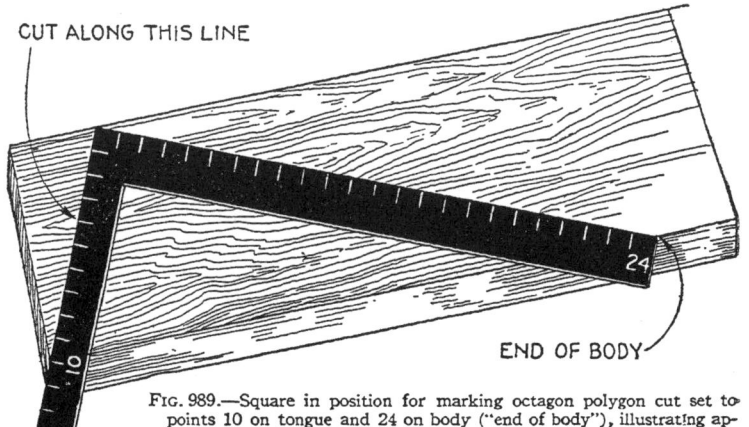

FIG. 989.—Square in position for marking octagon polygon cut set to points 10 on tongue and 24 on body ("end of body"), illustrating application of the reading shown in fig. 988.

regular rafter tables just described and need no further explanation except the last line or bottom row of figures which gives the bevel of intersecting lines of various regular polygons. It is used as follows: At the right end of body on the bottom line may be read *mitre cuts for polygons—use end of body*.

Example.—Find angle cut for an octagon.

For a figure of 8 sides to the right of the word *Oct.* in last line of figures find 10. This is the tongue reading, the end of the body being the other point as shown in fig. 989.

44 How to Use the Steel Square

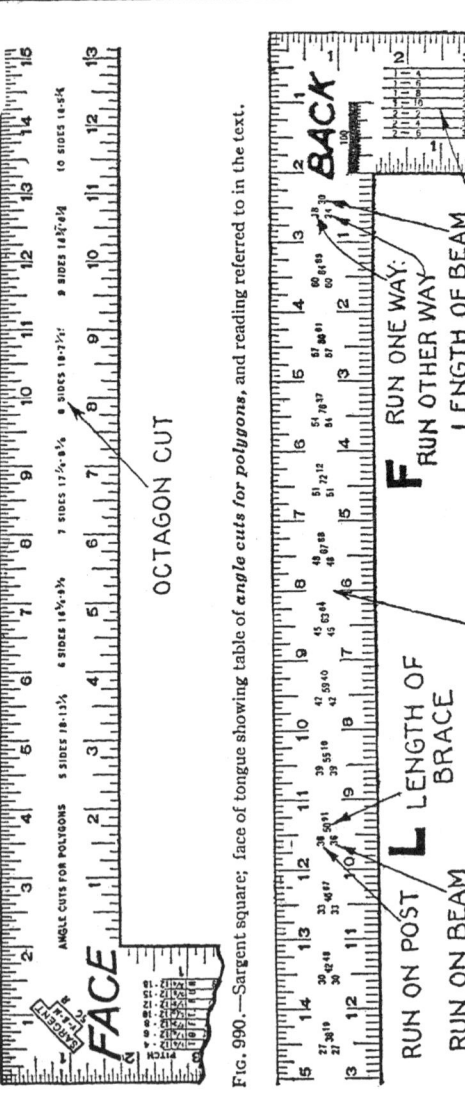

FIG. 990.—Sargent square; face of tongue showing table of *angle cuts for polygons*, and reading referred to in the text.

FIG. 991.—Sargent square; back of tongue showing table of *brace measure* and reading referred to in the text.

Table of Angle Cuts for Polygons.—On the Sargent square this table is found on the face of the tongue and gives setting points at which the square should be placed to mark cuts for common polygons having from 5 to 12 sides.

Example. Find bevel cuts for an octagon or 8 side.

On the face of the tongue (fig. 990) look along line marked "Angle cuts for polygons" and find the

How to Use the Steel Square

reading "8 sides 18—7½." This means that the square must be placed at 18 on one arm and 7½ on the other to obtain the octagon cut as in fig. 992.

Table of Brace Measure.—This table on the Sargent square as shown in fig. 991 is along the center of the back of the tongue and gives the length of common braces.

Example.—If the run be 36 ins. on the post and the same on the beam, what is the length of the brace?

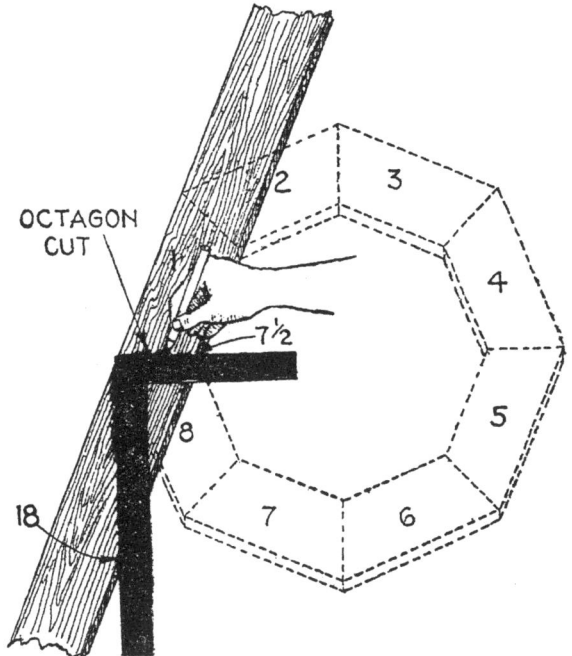

FIG. 992.—Application of square for angle cuts of polygons. The square is here shown set to points 18 and 7½ In constructing an 8-sided figure as an octagon cap for instance, mark the last figure in the reading is the setting for marking side; mark as shown Saw eight pieces of equal length, having this angle cut at each end of each piece, and the pieces will fit together to make an eight-sided figure, in size depending upon the length of the pieces. The dotted lines show figure as it would appear with the eight pieces in position.

How to Use the Steel Square

In the brace table or collection of figures, along the central portion of the back of the tongue (figs. 991 and 993) look (at L) for

$$\frac{36}{36}\ 50.91$$

This reading means that for a run of 36 ins. on post and 36 ins. on beam the length of beam is 50.91 ins. At the end of the table (at F near body) will be found the reading

$$\frac{18}{24}\ 30$$

This means that where the run is 18 ins. one way and 24 the other, the length of brace is 30 ins.

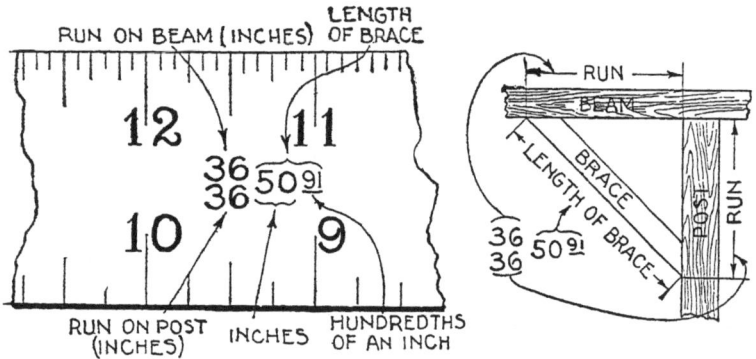

FIG. 993.—Portion of brace measure table with explanation of the various figures.

FIG. 994.—Brace in position illustrating measurements and reading of brace measure table.

The best way to find length of brace for runs not given on square is to multiply length of run by 1.4142 ft. (when run is given in feet) or by 16.97 ins. (when run is given in inches). This rule applies only when both runs are the same.

Octagon Table or Eight Square Scale.—This table on the Sargent square is located along the middle of the face of the

---

NOTE.—*Hundredth scale.* On most squares there is a scale of 1 inch length graduated into 100 parts, sub-divided into 20 parts, that will be found convenient when using the brace and rafter tables where decimal fractions occur. This is usually located in the corner near the brace measure table.

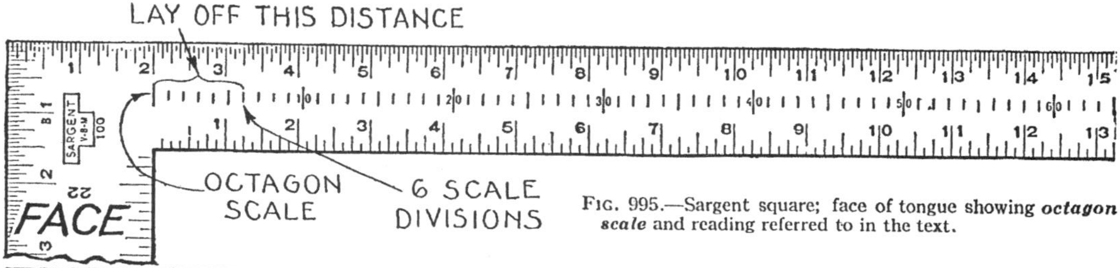

How to Use the Steel Square

Fig. 995.—Sargent square; face of tongue showing *octagon scale* and reading referred to in the text.

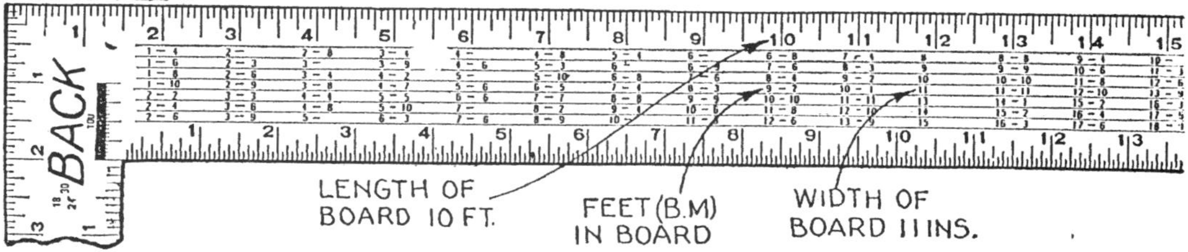

Fig. 996.—Sargent square; back of tongue showing table of *Essex board measure* and reading referred to in the text.

tongue and is used for laying off lines to cut an *eight square* or octagon stick of timber from a square timber.

In fig. 997, let ABCD, represent the end section or butt of a square stick of timber 6 × 6 inches. Through the center draw the lines AB and CD, parallel with the sides and at right angles to each other. With dividers take as many spaces, 6, from the scale as there are inches in width of the stick, and lay off this space on either side of the point A, as A*a* and A*h*; lay off in the same way the same spaces from the point B, as B*d*, B*e*; also C*b*, C*c*, and D*f*, D*g*. Then draw the lines *ab*, *cd*, *ef*, and *gh*. Cut off at the edges to lines *ab*, *cd*, *ef*, and *gh*, thus obtaining the octagon or 8-sided piece as in fig. 998.

Essex Board Measure Table.—This table is shown in fig. 996, and appears on the back of the tongue on the Sargent square. In applying the table the inch graduations on the outer edge of the square are used in combination with the values along the five parallel lines.

After measuring the length and width of the board, look under 12 in. mark for the width in inches. Then follow the line on which this width is stamped toward either end until the inch marked is reached on the edge of

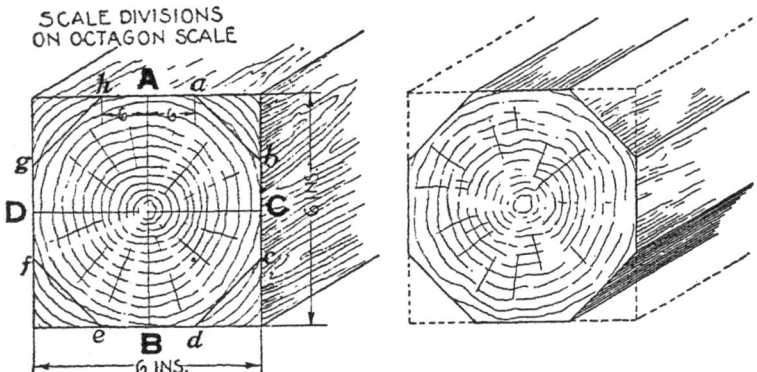

FIGS. 997 and 998.—Square timber and appearance after being cut to octagon shape, showing application of octagon scale in laying sides of an octagon or eight-sided polygon.

the square where number corresponds with the length of the board in feet, and the number found under that inch mark on the line followed will be the feet and inches contained in the board. The first number is feet and second number inches.

Instead of a dash between the ft. and ins. numbers, some squares have the inch division continued across the several parallel lines of the scale appearing on one side of the vertical inch division lines and ins. on the other.

Example.—How many feet Essex board measure in a board 11 ins. wide 10 ft. long and 1 in. thick? 3 ins. thick? Under the 12 in. mark on the outer edge of the square (fig. 996) find 11, which represents the width of the board in ins. Then follow on that line to the 10 in. mark (representing length of board in feet) and find on line 9, 2 which means that the board contains 9 ft. 2 ins. board measure for thickness of 1 inch. If the thickness were 3 ins. then the board would contain (9 ft. 2 ins.) ×3=27 ft. 6 ins. B.M.

Index

Angle table for square 9
Angles, laying off, 30° and 60° 8
Angles, laying off, 45° 9
Angles, laying off, any size 9
Bevel, for use with square for calculation..12
Brace cuts markings 3
Circle, of area equal to square of given size,
    to find diameter of 13
Circle, with area equal to sum of areas of two
    given circles, to find diameter of 7
Circle, circumscribing, to find radius from
    given side of polygon 15
Circle, to describe through three points not in
    a straight line 7
Circle, to divide circumference into given
    number of equal parts 15
Circle, to find circumference from diameter 12
Circle, to find the center 7
Circumference of a circle, to find from diameter 12
Circumference of circle, to divide into given
    number of equal parts 15
Circumscribing circle, to find radius from given
    side of polygon 15
Cripple rafters 24
Diagonal of a square, to find 12

Diameter of circle of area equal to square of given size	13
Eagle square	38
Eagle square, illustrated	41
Essex board measure table, use described	48
Essex markings	3
Framing square, defined	1
Framing square, parts	1
Gear wheel, calculation of pitch and number of teeth	13
Heel cut, defined	18
Heel cut, illustrated	19
Hip rafters	22
Jack rafters	24
Length of rafter, real and artificial defined and illustrated	18, 19, 20
Markings, defined (scales, tables)	3
Molding cuts	42
Number of teeth in gear wheel from pitch and diameter	13
Octagon cuts markings	3
Octagon inscribed in square	26
Octagon polygon cuts on Eagle square	42
Octagon rafters, table	42
Octagon, finding for any size of square timber	9
Pitch of gear wheel from diameter and number of teeth	13
Pitch table	20
Pitch, full, defined	19
Pitch, obtaining with framing square	21
Plancier cuts	42
Plate, illustrated	16

Plate, defined 17
Plumb cut, defined..18
Plumb cut, illustrated...19
Polygons, angle cuts for, illustrated 45
Polygons, angle cuts for, table 44
Radius of circle, to find, given length of chord 16
Rafter lengths, finding without use of tables 24
Rafter markings 3
Rafter table, common, on Sargent square 30
Rafter table, hip or valley, on Eagle square 41
Rafter table, hip, on Sargent square 32
Rafter table, jack, on Eagle square..41
Rafter table, jack, on Sargent square 35
Rafter table, main, on Eagle square 41
Rafter tables, described 29
Rafter, hip, cuts illustrated 37
Rafter, jack, cuts illustrated 36
Rafter, jack, side and bottom cuts illustrated 34
Rafter, length per foot run...38
Rafters, illustrated (common, hip, valley,
 jack, cripple) 16
Ridge timber, defined 17
Rise, defined 17
Run, defined 17
Sargent square, brace table..44, 45
Sargent square, Essex board measure table 44, 47
Sargent square, octagon table..46
Scale problems 3
Semi-circle, describe with given diameter 5
Side of square of area equal to circle of given
 diameter 12

Southington standard and take-down square, illustration	14
Southington Hardware Co. square, described	38
Southington Hardware Co. square, illustrated	40
Square and bevel problems	10
Square as a calculating machine	10
Square points for top and bottom cuts	29
Square timber, finding octagon	9
Square, cheap, markings	3
Square, markings, complete	3
Square, Sargent, described	30
Square, Sargent, illustrated	31
Square, to find the diagonal	12
Stanley square	38
Valley rafters	23